Reflexivity in Social Research

Emilie Morwenna Whitaker · Paul Atkinson

Reflexivity in Social Research

palgrave
macmillan

Emilie Morwenna Whitaker
University of Salford
Manchester, UK

Paul Atkinson
Cardiff University
Cardiff, UK

ISBN 978-3-030-84094-5 ISBN 978-3-030-84095-2 (eBook)
https://doi.org/10.1007/978-3-030-84095-2

This Palgrave Macmillan imprint is published by the registered company Springer Nature Switzerland AG
The registered company address is: Gewerbestrasse 11, 6330 Cham, Switzerland

Contents

1

Varieties of Reflexivity

Abstract There are numerous discussions of reflexivity and social research. The literature is, however, characterised by a variety of usages. Some are contradictory, and there is a state of confusion. The chapter outlines and examines some of those competing approaches, and suggests that they often do not go far enough or deeply enough into the full ramifications of research reflexivity. We introduce the main theme of the book: that reflexivity is a pervasive feature of all research, not just in the social sciences. It refers to the inescapable fact that the phenomena described in research are partly constructed by the methods used to study them.

Keywords Reflexivity · Social research · Methodology · Qualitative research

Introduction

It is not hard to find references to reflexivity among contemporary social sciences—especially but not exclusively in texts of qualitative research.

E. M. Whitaker and P. Atkinson, *Reflexivity in Social Research*,
https://doi.org/10.1007/978-3-030-84095-2_1

Indeed, in recent years it has become especially visible in the methodological literature and in the reporting of empirical studies. In some ways, that tendency reflects a very welcome degree of interest in the relevant issues for social research, and increased attention to topics of epistemology and methodology. On the other hand, the sheer volume of discussions of reflexivity includes a multiplicity of meanings, often inconsistent. At the very least, contemporary students, teachers and researchers are liable to find some difficulty in picking their way through the different formulations. The net effect is, certainly, that an appeal to reflexivity on the part of an author does not have a self-evident meaning or value for readers or practitioners.

In the course of this brief overview, therefore, it is our intention to try to make sense of that variety and those potential confusions. As will become apparent, we do not try to reduce the many connotations of reflexivity to a single 'right' formula. Equally, however, we do believe that there are weaker, and sometimes misleading, usages attached to the term. Hence our task here is to provide some systematic guidance as to what reflexivity can mean, sometimes what we think it therefore should not be taken to mean, and what the implications are for social research and its conceptualisation. We try to avoid imposing one unduly prescriptive approach, while advocating our particular perspective on the relevant methodological issues. We are concerned primarily with issues of research methodology, rather than philosophical problems. The latter are, of course, important, but really demand extensive separate treatment of some complexity.

In the course of this discussion, we shall, therefore, seek to clarify what reflexivity can mean, and consider its implications for the conduct and scrutiny of social research. We begin by outlining some of the variety and confusion surrounding the term. We shall then go on to discuss the generic issue of *epistemic reflexivity*. That is, reflexivity as a pervasive and generic issue in all research and in the social sciences most importantly. We shall then discuss selected aspects in more detail. *Disciplinary reflexivity* refers to the ways in which the disciplines, such as anthropology or sociology, frame what are and are not 'counted' as research topics and methods; *methodological reflexivity* continues in that vein, arguing that research methods are not simply 'the right tools for the job'. Chosen research methods actively constitute the kinds of phenomena that they describe. The description of phenomena then leads us to consider *textual* reflexivity, insofar as our forms of written representation also constitute the phenomena they account for. We then go on to discuss *positional reflexivity* and *reflexivity of membership* where both concern the methodological implications of the researcher's own biography and identity. We conclude that reflexivity is not a problem to be 'solved', nor is it a virtue to be celebrated. Hence we examine the broad implications for research conduct returning to the issues of 'objectivity' and 'subjectivity' that are raised in our earlier sections.

As we have suggested, it is important to seek clarity regarding reflexivity because it is invoked so frequently in published reports of research. One repeatedly reads claims that 'we conducted a reflexive ethnography', or advocacy of 'reflexive methods', or words to that effect. It is implied that reflexivity is a matter of choice, and that it is a virtue on the author's part. It is used in rather vague ways and is too often written about as an individual attribute of researchers, especially those conducting qualitative work. It is used to imply a degree of self-consciousness or self-awareness on the part of researchers and authors. While a recognition of research reflexivity does imply a degree of scrutiny and self-scrutiny, that by no means exhausts the methodological implications and significance of reflexivity in general. Consequently, our aim in this book is to provide a systematic account of reflexivity, as a corrective to unduly simple understandings.

Several Types of Reflexivity

Our section subtitle is a conscious allusion to Empson's *Seven Types of Ambiguity* (1930). Empson's exercise has nothing to do with resolving ambiguities: it explores the possibilities of ambiguity for the interpretation of texts (mostly English verse). In the same spirit, we do not attempt to resolve—far less eliminate—the implications of reflexivity. Rather, the intellectual task for all social researchers is to comprehend and work through the unavoidable reflexivity of their work. An important step in that intellectual process is also the recognition that 'reflexive' research is not a matter of virtue signalling, nor of purely personal choices. The issues of reflexivity are pervasive. It is necessary to recognise that there is no perfect or virtuous standpoint from which to undertake social research, and there is no research method that uniquely and transparently captures social realities. We cannot write reflexivity out of our accounts, even if we wanted to.

We can illustrate some of the variety of usage from published books and papers, drawn almost at random. So here are some examples:

> I conceptualize queer reflexivity as a practice that entails reflecting on how we shift into and out of the closet in multiple ways over the course of the research process through interactions with others, as well as the consequences of this shifting for the research process and relationships with participants. (McDonald, 2016: 392)

> If we argue that the activities and texts of our informants are really expressing not their obvious surface message but an underlying one about the nature of their society, then, in a reflexive displacement of this analysis, we may question the researcher's (our own) activities in producing a text about those others. (Davies, 2008: 9)

> ...the importance of *being* reflexive is acknowledged within social science research and there is widespread recognition that the interpretation of data is a reflexive exercise through which meanings are made rather than found. (Mauthner & Doucet, 2003: 414)

> Despite the extant advice available on interviewing elites, few scholars have engaged with the usefulness of reflexivity as a tool to assist in illuminating the dynamic nature of identity in the elite interview. (Mason-Bish, 2019: 264)

> We intended our approach to TA [Thematic Analysis] to reflect our view of qualitative research as creative, reflexive and subjective, with researcher subjectivity understood as a resource ... rather than a potential threat to knowledge production. (Braun & Clarke, 2019: 3)

> The ability to put aside personal feelings and preconceptions is more a function of how reflexive one is rather than how objective one is because it is not possible for researchers to set aside things about which they are not aware. (Ahern, 1999: 408)

> I argue that reflexivity focuses on the unfamiliar, the uncomfortable, the messy, difference/s, and writing up our failures.... (Lumsden, 2019: 4)

> Reflexivity is fundamental to an embodied process of discovery. Reflexivity is closely linked to positionality. In order to understand and process the information we have, we need to be aware of who we are, where we have come from, and how that is influencing our understanding.... By foregrounding both positionality and reflexive processes, we are able to be authentic to our selves and our experiences. (Leigh, 2021: 74)

We could go on reproducing similar comments about reflexivity, and as this book unfolds we shall obviously have occasion to cite many more. But from this sample alone, one can see that there are loose family resemblances between those propositions. They seem mostly to imply that reflexivity is a methodological or personal choice, that it complicates how research is conducted and conceptualised, and that it has something to do with personal, or even subjective justifications for social research. It also has something to do with critical, self-conscious 'reflection' about one's self as a researcher, and so results in 'authenticity' to selves and experiences. It can therefore range in use from an inherent feature of all research processes, to being a methodological choice, to a matter of personal identity. All of these and more will require unpicking in the course of this book.

Many references to reflexivity appear to be symbolic or ritualistic, rather than based on thorough examinations of what reflexivity can mean for research and how we conceptualise it. Macfarlane (2021) lists 'reflexivity' among the ideas often appealed to by qualitative researchers in a ritualistic way. He makes the strong claim that many accounts of research practice amount to academic 'faking it': 'There is an increasing tendency ... for the philosophy and language of close-up, qualitative research – criticality, reflexivity, statements of positionality, discussions of insiderism, and so on—to be the subject of a strategic deception'. We ourselves do not mean to imply that all appeals to issues such as reflexivity or positionality are intentionally 'fake', or that they are all made in bad faith. Macfarlane over-states his case. We do, however, agree with Macfarlane that far too many students and early-career researchers are encouraged to adopt and express a variety of philosophical and methodological 'positions' that include claims to have conducted 'reflexive' research, with little exploration of what that might mean in general, or what the full implications are for their own research in particular. In the course of doctoral supervision and more general mentoring, we detect pressure on graduate students to articulate a range of such positions, relating to choice of methods, procedures of data analysis, styles of textual representation and so on. Expectations on the part of examiners and peer reviewers can amplify such pressures, so that symbolic appeals to matters like reflexivity become standardised in the relevant discussions of qualitative method. Macfarlane consciously echoes Janesick, who observed the widespread ritualised appeals to methodological issues and the problems of 'methodolatry' (Janesick, 1994). Methodolatry implies the ritualised acknowledgement of epistemological and methodological literature, accompanied by citations to canonical authors. Reflexivity is one aspect of the catechism of methodological correctness, often found in the 'methods' sections of papers and dissertations.

This quasi-ritualistic trotting out of 'reflexivity' as a personal choice is part of the muddle surrounding the topic. These personal understandings of reflexivity tend to be poorly grounded in the epistemological foundations of sociological and anthropological thought. Rather than being a matter of collective research practice, this latter (mis)use stresses personal methodological choice. Reflexivity is portrayed as a matter of personal reflection and interpersonal sensitivity on the part of the researcher. It thus becomes a site of implied self-congratulation and self-regard, rather than a pervasive feature of the entire research process. We shall, therefore, try to dispel the misapprehension that reflexivity (as a condition of all research) is equivalent to reflection or reflective practice (as an individual research virtue), and insist that it certainly should *not* be equated with introspection or autobiographical 'confession'. We recommend that researchers should engage in reflective practice (Schön, 1983), through a critical examination of their research methods and preconceptions, and we expand on that later in this book.

Anthropologists have been especially prominent in discussions of reflexivity. Reflexive ethnography, as discussed by Davies (2008), 'expresses researchers' awareness of their necessary connection to the research situation and hence their effects upon it' (p. 7). In other words, it demands comprehension of the inevitable complexity of relations *in* the field, and the researcher's relationship *to* the field. Reflexivity in anthropology leads a consideration of the conduct of fieldwork, particularly the anthropologist's identity and relationships with the 'others' with whom they work. It also reflects back on the ethnographer's relations with the discipline of anthropology itself. Reflexivity, therefore, implies the realisation that 'any statement about culture is also a statement about anthropology' (Crick, 1982: 307). Such a statement speaks to the distinctive nature of ethnographic practice. Members of the host society are engaged in the performance of their own culture. The reflexive nature of observation and participation rests on the extent to which social actors are themselves active in constructing their own versions of social reality. Ethnographic observation are not performed on otherwise inert objects. Social and cultural phenomena are produced through actors' everyday work of action and interpretation. Consequently, we shall need to pay attention to a principle of reflexivity that reflects the intensely social and dialogic foundations of ethnographic work. As a number of authors have suggested, performances such as rituals and other collective observances may be as much reflexive constructions of the field and of the culture as the work of the ethnographer herself (Myerhoff & Ruby, 1982). There is, therefore, a double process at work—as the ethnographer reflexively constructs cultural categories that are simultaneously being constructed by the participants themselves. Geertz's interpretation of the Balinese cockfight is often invoked in this context. Such reflexive enactments are then, of course, available for further textual or performative reconstructions (Schechner, 1982; Turner, 1982).

The general principle of reflexivity of research is by no means confined to the social sciences. Indeed, it is most clearly visible in the natural sciences. In order to grasp its significance we need to suspend some taken-for-granted, even lazy, perceptions of the sciences. A science such as quantum physics is entirely suffused with epistemic reflexivity, and that is explicitly recognised by the scientists themselves. Scientific observation is, as Barad (2007) illuminates (building on the pioneering physicist Nils Bohr), thoroughly dependent on 'apparatus', and apparatus includes all of the material, personal, interpersonal and institutional means necessary to render a phenomenon visible, measurable and amenable to manipulation. To some degree, all observable phenomena depend on the methods used to make them observable. Quantum-level physics displays that most forcibly, but the general argument extends to any kind of 'data' and 'measurement'. The same is true of the social sciences, although the notion of 'apparatus' should not be interpreted in too literal or simplistic a manner. In the social sciences, that apparatus may include specific techniques of data collection (interview schedules, questionnaires, personality inventories). It also involves the enrolment of research assistants, whose activities and interpretations are socialised in accordance with the research project's research strategies and assumptions. In ethnographic work it also includes the researcher her/himself as the primary form of 'apparatus', and therefore as the main source of observations. The assemblage that we can gloss as 'apparatus' also includes the enrolment of participants ('human subjects') who self-evidently exert their own subjectivity and agency. The apparatus of research in the social sciences includes all those means of 'data' collection, from the strategies of participant observation, to varieties of interview, to the collection of permanent recordings of everyday activity. All of these elements are intertwined in a network of observation, interaction and inference that yield the phenomena of the social sciences. In other words, we need to think of an array of actors, resources and techniques whereby everyday social life is rendered into 'observations', 'data', 'findings' and are then further transformed into written or other forms of representation (such as academic papers or books).

Quantum physics may seem a somewhat esoteric allusion, and comparisons with the natural sciences are often misleading. But the analogy with physics serves to underline the fact that all of the sciences are founded on some common principles (that are not to be equated with the crudest versions of positivism) that are not just matters of choice for the individual researcher. All observations and descriptions are, to some degree, shaped by the 'apparatus' that is mobilised in order to render them observable and describable. One cannot, from this perspective, *choose* to be reflexive, much less choose to avoid reflexivity. The act of observation is itself reflexive, whether one likes it or not, whether one recognises it or not. Consequently, the methodological task is to identify *how* reflexivity is to be understood, and what the consequences—actual and potential—are for social research in general, and for qualitative social science in particular. As Hammersley and Atkinson (1997) make clear, that does not mean that we can or should abandon any claims to be able to represent the social world: '…to say that our findings, and even our data, are *constructed* does not automatically imply that they do not or cannot represent social phenomena' (p. 16, emphasis in original). A belief in the unmediated perception and comprehension of the social world would be to assume 'that the only true form of representation would involve the world imprinting its characteristics on our senses without any activity on our part' (ibid.). We shall reiterate this important assertion in a later section. But for now we wish to introduce the fundamental point: that to assert the constructed nature of research does not mean that it is arbitrary, 'subjective', or spurious.

Although it features most prominently in the methodological literature of sociology and anthropology, reflexivity, from our perspective, is a fundamental feature of all science. The most foundational of the sciences—physics—in fact provides a key example. As we have already suggested, the most elementary of the sciences—quantum physics—currently includes the principle of reflexivity (though not by that term) and has done so since the revolutionary science of Einstein, Bohr, Heisenberg, Dirac and others. It proposes that it is the act of observation itself that creates the state of a particle, which is in a state of 'superpositionality' until the observation is made, and the 'wave function' collapses. As Barad (2007) has explored, physics and the natural sciences are thoroughly characterised by multiple entanglements. Images of detached observers engaging with inert materials are far from accurate representations of the natural sciences, and so crude distinctions between the 'natural' and 'cultural' disciplines are misplaced.

As we shall explain, Barad's own representation of 'reflexivity 'in the social sciences is potentially misleading, however, and so while her discussion of quantum mechanics is valuable, her summaries of social science do not do full justice to social scientists' explorations of the core issues. It seems to derive from a very restricted reading of the social-science understandings of reflexivity. Consequently, her contribution to a collective understanding of reflexivity is frustrating. Given the prominence of Barad's work, it is worth outlining her contribution and seeking to clarify what we see as a confusion.

Barad writes explicitly about 'reflexivity' only to declare it epistemologically inadequate:

> Reflexivity, like reflection, still holds the world at a distance. It cannot provide a way across the social constructivist's allegedly unbridgeable epistemological gap between knower and known, for reflexivity is nothing more than iterative mimesis: even in attempts to put the investigative subject back in the picture, reflexivity does nothing more than mirror mirroring. Representation raised to the nth power does not disrupt the geometry that holds object and subject at a distance as the very condition for knowledge's possibility. Mirrors upon mirrors, reflexivity entails the same old geometrical optics of reflection. (Barad, 2007: 87–88)

The problem here, as we see it, is that—derived from relatively few sources—Barad thinks that sociological treatments of reflexivity are based only on matters of 'representation', such as the textual conventions of academic writing. In other words, citing authors like Woolgar (1988), she assumes that reflexivity is primarily an issue of what we shall describe as textual reflexivity. In one way that reading of Woolgar is understandable. He seems to begin with a generic and programmatic statement of reflexivity: 'The very attributes of nature, the way in which the physical world is apprehended, described and classified, depend on the technologies which make these activities possible' (p. 88). That sounds pretty comprehensive. But then his discussion does switch to an emphasis on 'representation' and in his concluding remarks he writes: 'We suggested …the development of an alternative, reflexive perspective on science which self-consciously takes representation as its topic'

(p. 94). Now that particular stress on 'representation' and textual strategies in particular, was a distinctive feature among some Science and Technology Studies scholars at the time Woolgar was writing. Reflexivity was discussed primarily in terms of textual practices and experiments. A notable example was the book by Malcolm Ashmore (1989), the very title of which conveyed a general approach, *The Reflexive Thesis: Wrighting Sociology of Scientific Knowledge*, whole others, such as Michael Mulkay were also experimenting with 'alternative' forms of reportage (e.g. Mulkay, 1985). Such approaches, which sought to disrupt taken-for-granted forms of academic practice, and paralleled the STS interest in rendering strange scientific practice itself, certainly tended to equate reflexivity with issues of representation. Like many other versions of reflexivity, that STS perspective also enjoined strategies of critical self-awareness on the part of the social scientist. Donna Haraway (1991) was critical, commenting on: '...Woolgar's relentless insistence on reflexivity, which seems not to be able to get beyond self-vision as the cure for self-invisibility' (p. 33). Again, we shall argue that a simple version of 'self-vision' is at best an incomplete version of reflexivity.

Barad and Haraway both prefer a different way of conceptualising the issues, using the metaphor of 'diffraction'. As we have seen, Barad seems to assume that discussions of reflexivity are based on notions of 'reflection' that in turn imply a mirror-like relationship with Nature. So she contrasts her use of Diffraction with Reflection. By diffraction she intends an understanding of inquiry based (a) on 'performativity' rather than 'representationalism', (b) emergence of subject and object through intra-actions, rather than a 'pre-existing boundary between subject and object' (c) entangled states of nature cultures, rather than a nature/culture binary (p. 89); see also Haraway (1991: 33–36). Now we see as much—if not more—of 'reflexivity' in Barad's 'diffraction' as her version of 'reflexivity'. On reflexivity and diffraction, along with other usages, see: Bozalek and Zembylas (2017) and Schneider (2002). The imagery of diffraction is among a number of such metaphors, as Lynch (2000) notes: 'Instrumental and optical metaphors abound in this context: the reflective and refraction processing of "reality"; the dependence of appearances on observational "standpoints"'... (p. 29). Now the purpose of this commentary on Karen Barad is not to single her out for gratuitous criticism. On the contrary, the substance of her book is a useful resource for anyone investigating these topics. But it does suggest—and the rest of this book will also suggest—that trying to encapsulate these complex and contested ideas under a single rubric, whatever it may be, is fraught. That is why the structure of this book tackles a discrete number of themes, breaking down the issues into tractable topics. We are, of course, stuck with the overall term of reflexivity, but our approach is intended to underline just how multi-faceted the ideas are in practice.

Our own perspective owes more than a little to the idea of the *reflexivity of accounts*. The means and devices that are used to produce an account or description of something contributes to and frames the phenomenon it describes. There is, in other words, a degree of circularity between the phenomena that research analyses and the means used to achieve or identify those same phenomena. We must emphasise—and will do so again—that this is not intended to lead to the sort of constructionism that implies that the objects of research and research methods are arbitrary. We stress the pragmatist research tradition in the sociology of scientific and social research: that it is grounded in practical, concrete engagements with the world about us. Understanding is forged through experience and exploration. The construction of reality reflects the extent to which such exploration depends on human activity, human judgement, and the everyday *work* of research itself. There are many means that are brought to bear on such practical action, including—but not restricted to—the methodological resources available to a research community at any given time, and within disciplinary boundaries. These are collective engagements, grounded in disciplinary and methodological understanding, and are not simply matters of personal preference.

Recognition of such collective action and its consequences does not in and of itself imply that research is irretrievably 'biased'. Consider, for instance, the example of the crash-test dummy. Such figures are 'accounts' or 'models' of the human figure. They reproduce key features of such bodies, while ignoring others. The features that are chosen inevitably impinge on the sorts of research measurements that can be derived from experimental car crashes, and the kinds of phenomena that can be studied. (They do not typically model internal organs for instance.)

Such dummies are, therefore, 'accounts' of bodies. On the other hand, the revelation that the physical dummies are based on *male* bodies means that they are very poor, if not useless, in modelling the effects of car crashes on women's bodies (Perez, 2019). The latter *is* a source of bias in the research, while the former reflects the ubiquitous features of any and every account or model. The generic aspect of models was discussed by Garfinkel and Sacks (1970), who describe them as 'glosses' on the phenomena they stand for. Now clearly, for practical purposes, any given model or gloss can be evaluated, as adequately faithful to desired features or otherwise, as adequately sensitive or detailed, and so on. The issue is not whether accounts or glosses or models are 'good' or 'bad', but *how* they reconstruct the phenomena they purport to stand for; see also Lewis et al. (2013) for a parallel discussion of animal models in biomedical research, where species of laboratory animals are selected to model specific features of clinical pathology, and therefore only have to be 'good enough' for practical research purposes.

It would, therefore, be wrong to associate the reflexive nature of research with bias or even error. It is possible that such differences can be attributed to erroneous interpretation: to the over-enthusiastic pursuit of an *idée fixe*, for instance, or to unacknowledged ideological assumptions. But it is equally clear that they can reflect distinctive epistemic conditions, as well as methodological and interpretative distinctions. They can reflect changing paradigmatic presuppositions, methodological inscriptions, historical intellectual preferences and national research traditions. The same considerations apply to agreement and stability in interpretations and measurements too. Consequently, discussions of reflexivity that are couched reductively in terms of bias are far too simplistic.

From the point of view we are exploring here, we should think always in terms of the essential reflexivity of methods, rather than 'reflexive' methods as constituting a particular class of research, or even a particular orientation towards the research process. This is where we part company with authors such as Alvesson and Sköldberg (2018), who focus exclusively on 'qualitative' research, and concentrate their attention on varieties of recent and contemporary perspectives, such as ethnomethodology, hermeneutics, critical theory, poststructuralism and postmodernism. This is a fairly familiar litany of epistemologies, that certainly can be held to challenge more 'traditional' approaches to empirical research. But it is our contention that to identify reflexivity closely with those strategies and perspectives is to miss the underlying point. It certainly overlooks the extent to which quantitative methods, positivist epistemologies and the like reflexively constitute the objects of their scrutiny. If we accept the basic premise that methods inscribe ideas about what might count as observable and researchable phenomena, then we cannot exempt some—albeit by implication—while apparently celebrating others for being 'reflexive' (or possibly more reflexive than others).

It should now be clear that the topic of reflexivity is pervasive in the methodological literature of the social sciences, and it is especially prominent in writing about qualitative research. We have also argued that reflexivity is an issue for the social sciences in general. It is indeed a feature of all research, across the natural, social and cultural disciplines. But it is clear that it has particular resonance and urgency for qualitative and ethnographic research. In the course of this discussion, therefore, we shall focus primarily on contemporary qualitative research. In the next chapter we shall discuss the general phenomenon of *epistemic reflexivity*. This is an overarching term that in principle encompasses all aspects of reflexivity in all disciplines. In subsequent sections we shall examine epistemic reflexivity in more detail, where we discuss the effects of academic discipline or field, the implications of research method, researchers' positionality and biography, and the reflexivity of representations.

References

Ahern, K. J. (1999). Ten tips for reflexive bracketing. *Qualitative Health Research, 9*(3), 407–411.

Alvesson, M., & Sköldberg, K. (2018). *Reflexive methodology: New vistas for qualitative research* (3rd ed.). Sage.

Ashmore, M. (1989). *The reflexive thesis: Wrighting sociology of scientific knowledge*. University of Chicago Press.

Barad, K. (2007). *Meeting the universe halfway: Quantum physics and the entanglement of matter and meaning*. Duke University Press.

Bozalek, V., & Zembylas, M. (2017). Diffraction or reflection? Sketching the contours of two methodologies in educational research. *International Journal of Qualitative Studies in Education, 30*(2), 111–127.
Braun, V., & Clarke, V. (2019). Reflecting on reflexive thematic analysis. *Qualitative Research in Sport, Execise and Health, 11*(4), 589–597.
Crick, M. (1982). Anthropology of knowledge. *Annual Review of Anthropology, 11*(1), 287–313.
Davies, C. A. (2008). *Reflexive ethnography: A guide to researching selves and others* (2nd ed.). Routledge.
Empson, W. (1930). *Seven types of ambiguity*. Chatto & Windus.
Garfinkel, H., & Sacks, H. (1970). Formal studies of practical action. In J. C. McKinney & E. A. Tiryakian (Eds.), *Theoretical sociology* (pp. 337–366). Appleton-Century-Crofts.
Hammersley, M., & Atkinson, P. (1997). *Ethnography: Principles in practice* (3rd ed.). Routledge.
Haraway, D. (1991). *Simians, cyborgs and women: The representation of nature*. Free Association Books.
Janesick, V. J. (1994). The dance of qualitative research design: Metaphor, methodolatry, and meaning. In N. K. Denzin & Y. S. Lincoln (Eds.), *Handbook of qualitative research* (pp. 209–219). Sage.
Leigh, J. (2021). What would a rhythmanalysis of a qualitative researcher's life look like? In B. C. Clift, J. Gore, S. Gustafsson, S. Bekker, I. C. Batlle, & J. Hatchard (Eds.), *Temporality in qualitative inquiry: Theories, methods and practices* (pp. 72–92). Routledge.
Lewis, J., Featherstone, K., & Atkinson, P. (2013). Representation and practical accomplishment in the laboratory: When is an animal model good-enough? *Sociology, 47*(4), 776–792.
Lumsden, K. (2019). *Reflexivity: Theory, method and practice*. Routledge.
Lynch, M. (2000). Against reflexivity as an academic virtue and source of privileged knowledge. *Theory, Culture & Society, 17*(3), 26–54.
Macfarlane, B. (2021). Methodology, fake learning, and emotional performativity. *ECNU Review of Education*, published online March 19, 2021.
Mason-Bish, H. (2019). The elite delusion: Reflexivity, identity and positionality in qualitative research. *Qualitative Research, 19*(3), 263–276.
Mauthner, N. S., & Doucet, A. (2003). Reflexive accounts and accounts of reflexivity in qualitative data analysis. *Sociology, 37*(3), 413–431.
McDonald, J. (2016). Expanding queer reflexivity: The closet as a guiding metaphor for reflexive practice. *Management Learning, 47*(4), 391–406.

Mulkay, M. (1985). *The word and the world: Explorations in the form of sociological analysis*. George Allen & Unwin.
Myerhoff, B., & Ruby, J. (1982). Introduction. In J. Ruby (Ed.), *A crack in the mirror: Reflexive perspectives in anthropology* (pp. 1–35). University of Pennsylvania Press.
Perez, C. C. (2019). *Invisible women: Exposing data bias in a world designed for men*. Chatto & Windus.
Schechner, R. (1982). Collective reflexivity: Restoration of behavior. In J. Ruby (Ed.), *A crack in the mirror: Reflexive perspectives in anthropology* (pp. 39–81). University of Pennsylvania Press.
Schneider, J. (2002). Reflexive/diffractive ethnography. *Cultural Studies ↔ Critical Methodologies, 2*(4), 460–482.
Schön, D. (1983). *The reflective practitioner: How professionals think in action*. Temple Smith.
Turner, V. (1982). Dramatic ritual/ritual drama: Performative and reflexive anthropology. In J. Ruby (Ed.), *A Crack in the mirror: Reflexive perspectives in anthropology* (pp. 83–97). University of Pennsylvania Press.
Woolgar, S. (1988). *Science: The very idea*. Ellis Horwood.

2

Epistemic and Disciplinary Reflexivity

Abstract This chapter takes the discussion further by examining the most fundamental and far-reaching meaning of reflexivity. Epistemic reflexivity is the generic term we use to describe the pervasive feature of all research: the extent to which the phenomena of research are shaped and framed by the disciplinary presuppositions and methodological prescriptions that are brought to bear. We illustrate this principle in a discussion of disciplinary reflexivity: how academic disciplines determine the proper objects of research, while excluding others from their canon.

Keywords Epistemic reflexivity · Disciplinary reflexivity · Sociology of knowledge · Measurement · Pierre Bourdieu

Epistemic Reflexivity

As we have already implied, we cannot begin with a single definition of reflexivity. As we make clear, it has multiple meanings and connotations, and although there are family resemblances, those meanings cannot be reduced to a simple summary statement or formula. What we are committed to however, is stating that reflexivity in social research is not

E. M. Whitaker and P. Atkinson, *Reflexivity in Social Research*,
https://doi.org/10.1007/978-3-030-84095-2_2

an option, nor is it a research virtue. It is a fundamental and inescapable feature of all research, including—especially—social research. All aspects of reflexivity are epistemic, in that it bears on the conditions of knowledge in the social sciences, and on the collective work that informs knowledge-production. Here we use epistemic reflexivity as a useful heuristic, an overarching term, that encompasses those more specific types—disciplinary, methodological, positional, textual—that we explore in more detail as this book unfolds. We do of course owe a debt to Bourdieu in shamelessly borrowing a term most associated with his work, and we will turn shortly to how we position ourselves in this regard. First, it is important to sketch our approach to reflexivity.

Reflexivity refers fundamentally to the fact that, across all the sciences, natural and cultural, the very act of observing or measuring constitutes the phenomena being described. That to 'do research' in part constructs or defines the phenomenon that is the object of that research. Of course, classifications, descriptions and measurements are not completely arbitrary. We do not claim that phenomena in the natural and social domains are totally constituted or determined by how we observe and describe them. But the objects of research are framed by the kinds of questions we can ask, the kinds of measurements we can make, and the kinds of descriptions that are available to us. Any act of observation or measurement is a form of intervention, there is no perfectly neutral vantage-point, and no transparent medium of description, that exempt the observer from some degree of reactivity. This is by no means confined to the social sciences, although social scientists may be especially aware of it, and may confront its possible consequences more overtly than many others.

An awareness of reflexivity does not result in a position of pure relativism or the strongest versions of constructivism. In other words, it does not deny the existence of a physical or social reality that is independent of our investigations. Rather, it states that any knowledge of the natural or social world is necessarily framed in accordance with our ideas and our methods.

This general epistemological point has a long pedigree. The sociology of knowledge for many years recognised that knowledge is perspectival, or relational. Mannheim's sociology of knowledge, for instance, recognises that '...the social qualities of the knower shape the characteristic of his/her thought, not only with regard to the genesis of ideas, but also concerning the form and contents as well as the formulation and intensity of experience' (Mendel, 2006: 31). Indeed, as May and Perry (2017) demonstrate, the general idea has deep roots in a number of philosophical traditions, But, as we shall elucidate, it is not just the 'social position' of the knower that shapes her or his thought; or rather, the idea of social position needs to be explored in detail. It encompasses the scholar's disciplinary membership, the methodological approach adopted, the forms of representation that are deployed, as well as the investigator's own biography and identity. All of these impinge—often tacitly—on the conceptualisation and conduct of research. While, as we have said, epistemic reflexivity shapes all inquiry, it is especially pertinent in a consideration of the social disciplines.

Despite its recent prominence, the general issues of epistemic reflexivity (if not in name) have long been recognised and have featured in important contributions to methodological discussion. A key reference point here is Aaron Cicourel's pioneering *Method and Measurement in Sociology* (1964). In the years following its publication, Cicourel's book was one of the most cited methodological texts and one of the most frequently required texts on methods reading-lists. If it no longer enjoys that degree of visibility, its key messages remain significant, and are most pertinent to contemporary ideas about reflexivity. For evaluations of Cicourel's contribution and its continuing relevance, fifty years after its publication, see Smith and Atkinson (2016). Cicourel's use of the term 'measurement' led to misunderstanding—not least by people who did not seem to have read the original work carefully. Cicourel's argument is that any attempt to categorise, classify or otherwise describe a social phenomenon is a matter of measurement. So contrary to vulgar mis-readings, Cicourel was not primarily engaged in a critique of quantitative social science per se. His arguments applied to any and all forms of classification. His was not a partisan advocacy of qualitative research. Rather, he sought to identify some of the ways in which research methods create categories, types and phenomena. They construct variables, often standardising measurements. While varieties of survey design or attitude scaling are among the most characteristic contexts for such work, qualitative research, with its codes and themes, strategies and negotiations, is equally pertinent. The question, as we see it, is not to decry all such analytic work. Cicourel's contribution was to address *how* such methodological activities incorporate assumptions about phenomena that need to be examined critically. Equally, we should not be importing unexamined assumptions about the nature of social life or social actors by incorporating them tacitly into our categorising and measuring practices. In other words, Cicourel pointed to the same generic phenomenon as do discussions of reflexivity: that our methods of inquiry help to constitute the phenomena that we study. Our methods of observation, description and classification reflexively determine what can be studied and how it should be investigated. This is at the heart of the epistemic reflexivity that we explore in this book. It is clear that in an age of 'big data', and in a world awash with measurements and evaluations, close attention to methods and their products remains an urgent need.

At this point it is worth pointing out why the term itself has come to stand for this intriguing relationship between method and phenomena. Reflexivity means that in some general sense something arcs back on itself. Research methods, therefore, shape what they describe, while the phenomena that are revealed in turn justify those self-same methods. There is a dialectical, circular relationship between methods and what they describe. Indeed, description here becomes prescription, in that methodological strategies determine what *should* be observed. It *is* a constructivist perspective on knowledge and its production, but as we shall see, this is not the sort of vulgar constructivism that implies that anything constructed is invalid, bogus, or even not 'scientific'. Rather, it is based on the recognition that there is a relationship between phenomena and how we can study them, and that it is a matter of *work* on the part of intellectuals. This is not a matter that is restricted to sociology, or to the social sciences more widely. It is, if we are to take it seriously, a feature of all scholarly inquiry.

Bourdieu and Beyond

Our introductory remarks showed how many contemporary appeals to reflexivity stress the personal identity of the researcher, critical reflection on her or his positionality, and the subjective aspects of research practice. It is a tendency towards the personalisation of research, the confessional revelation of the author's biography, and the expression of personal, even emotional, engagements with the research. Hence some versions of 'reflexivity' insist on the *subjective* aspects of research and the interpretation of research materials.

On the other hand, there are, as we shall see, significant versions of reflexivity that portray it as a means towards greater *objectivity* in social research. For instance, it is invoked in the interests of making possible a 'science' of sociology. The following characterisation of Pierre Bourdieu's approach summarises it succinctly:

> His [Bourdieu's] analysis of intellectuals and of the objectifying gaze of sociology, in particular, like his dissection of language as an instrument and arena of social power, imply very directly, and in turn rest upon, a self-analysis of the sociologist as cultural producer and a reflection on the socio-historical conditions of possibility of a science of society. (Bourdieu & Wacquant, 1992: 36)

That summary statement brings together several key strands: reflexivity and the work of intellectuals, the power of analysts, and self-analysis on the part of the sociologist. Bourdieu's is a significant contribution, and we shall discuss it further below. The self-analysis he advocates is not simply a matter of introspection or an emphasis on the 'personal' aspects of research. And it is certainly not a celebration of the 'subjective' aspects of social research. It is about the collective work of the academic discipline or 'field'. For now, we note the apparent paradox: that reflexivity enhances the subjective aspects of research while also contributing to a more objective social science.

All aspects of reflexivity are epistemic, in that it bears on the conditions of knowledge in the social sciences, and on the collective work that informs knowledge-production. Epistemic reflexivity is the generic term that in turn captures those more specific types (disciplinary, methodological, positional, textual). In general terms, reflexivity refers fundamentally to the fact that, across all the sciences, natural and cultural, the very act of observing or measuring constitutes the phenomena being described. We shall not be claiming that phenomena in the natural and social domains are totally constituted or determined by how we observe and describe them. Classifications, descriptions and measurements are not completely arbitrary. But the objects of research are framed by the kinds of questions we can ask, the kinds of measurements we can make, and the kinds of descriptions that are available to us. So there is something important to state here about the role of the collective in shaping these features of social scientific practice. An awareness of reflexivity does not result in a position of pure relativism or the strongest versions of constructivism. In other words, it does not deny the existence of a physical or social reality that is independent of our investigations. Rather, it states that any knowledge of the natural or social world is necessarily framed in accordance with our ideas and our methods.

As we have begun to sketch, the genus 'reflexivity' now contains many different assumptions and prescriptions, often associated with a set of progenitors: Schutz, Garfinkel, Mannheim, or more contemporaneously the authors of methods textbooks and musings on social scientific research ethics. This is impressive given that 30-odd years ago, an author of a volume on reflexivity was able to proclaim the concept was, 'ignored, evaded, diminished' by most social scientists (Woolgar & Ashmore, 1988: 2). Underlying this diversity, however, is a thin understanding of reflexivity that amounts to a general argument that researchers should explicitly position themselves in relation to their objects of study so that an assessment can be made of their knowledge claims vis-à-vis situated aspects of their social selves. We address these issues of membership and naïve treatments of identity, later in the book. Putting aside how practicable or possible this kind of identity-diagnosis work is, the question is lost as to how reflexivity can be organised, 'accomplished' or enacted in research practice. In order to make our way through this muddle, we borrow briefly from Bourdieu and his concept of 'epistemic reflexivity'. To be clear, we are neither Bourdieusians, nor do we use this concept in a purist (or puritanical) way. Where we align with Bourdieu is on the importance of re-rooting reflexivity within a collective frame. We also share his concern with research practice, as Wacquant went on to state, 'Bourdieu's concern for reflexivity, like his social theory, is neither egocentric nor logocentric but quintessentially embedded in, and turned toward, scientific practice' (Bourdieu & Wacquant, 1992: 46).

Bourdieu's ideas on epistemic reflexivity (Bourdieu, 1990a, 1990b; Bourdieu & Wacquant, 1992) were published at a juncture—after the so-called 'crisis' of the late 1980s to 1990s over the status of sociological knowledge in general, and the performative nature of ethnographic writing in particular. His concept of epistemic reflexivity sought to get beyond a neopositivist camp without falling into poststructuralist relativism (cf Maton, 2003). Bourdieu set out to radicalise reflexivity in so far as he sought to move it to the collective level of concern, a kind of call or invitation for scholars specifically, to 'arc back' upon their disciplinary homelands: see Wacquant (1989) and Hess (2011). For him, there is a deeper 'scientific unconscious embedded in theories, problems, and (especially national) categories of scholarly judgement' (Bourdieu & Wacquant, 1992: 40) which needs to be rendered explicit. He referred to such analysis and excavation as 'objectifying objectification' on a collective basis (Bourdieu & Wacquant, 1992: 71–72). It is not merely the individual researcher who is of interest to Bourdieu here but rather the intellectual field. The aim is not to unveil an individual researcher on spurious biographical grounds, but to draw attention to the collective scientific unconscious embedded in intellectual practices, predilections and preoccupations. Plumbing the waters of the 'personal', which is where reflexivity often begins and ends, does not address those features. Personal confession or disclosure is not good enough in and of itself.

We align our argument with Bourdieu here too: he situates his work as a collective endeavour, undertaken by the social scientific field as a whole (Swartz, 1997). Epistemic reflexivity, as a shared project, invites the sociological community to become conscious of (and thus act on) the shared conditions, assumptions, expectations, framings with which we engage in constructing our sociological objects, our methods, our textual representations. Hence, as Maton (2003) says in relation to Bourdieu, '...both the object and the subject of reflexivity are collective (the intellectual field as a whole) rather than individual, and this collective reflexive analysis of collective objectifying relations will, Bourdieu argues, provide an epistemological basis for social scientific knowledge' (p. 58). As we shall see, it is not necessary to endorse Bourdieu's specific argument in order to endorse the view that reflexivity ought to be regarded as a collective matter, not one of individual orientation and choice. Moreover, quite how, given the varieties and flavours of sociology, members are to reach consensus about the terms for such an exercise remains unclear, however.

Epistemic reflexivity is committed to the analysis of the evolution of the object of research both within the social field where it is encountered by the researcher, and, within the academic field where it is conceptualised. Here, there are resemblances to the kinds of (in)famous 'history of the present' work undertaken by Foucault (Garland, 2014). Other more esoteric examples help us to think through these ideas of collective reflexivity Bourdieu was trying to establish. There is the political scientist, Reinhart Koselleck, who wrote an interpretation of the emergence of modernity exploring the role of intellectual clubs as his doctoral thesis (Koselleck, 1998 [1959]). The historian Norman Cohn, writing after the second world war, sought to dig out and bring to light the joint intellectual origins of totalitarian mass movements and the idea of progress from his own personal experience (Cohn, 1970 [1957]). All were concerned with how ideas and practices are accomplished and sedimented, how vestiges of past constructions persist, underpinning and informing what is emerging in the current present. These are all 'conscious' intellectual projects. They arc back upon their subject matter, their fields, their intellectual networks, excavating, mapping, shining lights on the 'apparatus' which guides and confines them. Perhaps of greater familiarity to the social sciences is the work of Norbert Elias, and his processual sociology. Szakolczai (2000) discusses the reflexivity of Elias's work, in an overview of Elias's intellectual formation, and the dense intersections of his intellectual biography and the historical sociology that provided the content of his career (see also Smith, 2000). Elias himself might well see the reification of personal and virtuous reflexivity as a 'naively egocentric' (Elias, 1970: 14) view of social life. He may well lament how at the very time 'reflexive researchers' seek to underscore their virtue as ethical agents practicing something called 'reflexivity', interdependencies are overlooked in the presentation of these projects.

Where we differ with Bourdieu, in a significant and important way, is his expectation that this reflexive work will render sociological knowledge more objective. His treatment of reflexivity is not intended to undermine the objectivity of sociology (and he deals only with sociology), but rather to enhance it. Bourdieu's project attempts to re-scientise sociology through an attention to epistemic reflexivity. Bourdieu argued that his conception of epistemic reflexivity provided not only a means of developing richer descriptions of the social world but also the basis for a more practically adequate and epistemologically secure social science. We disagree. We do not see reflexivity as either a maze to escape from or a toolkit to progress with to reach the possibility of a 'naturalistic' social science. Where Bourdieu takes Bachelard's 'applied rationalism' from the realm of the natural sciences to the realm of the social sciences (Vandenberghe, 1999), we see the issue of reflexivity as inevitably present across all scientific endeavours.

We subscribe to the collective call which Bourdieu was attempting to make—that all researchers need to engage in the work of reflexivity. Where we do profoundly differ is in our underscoring of reflexivity as something we 'live with' rather than something to escape from or as a tool to better 'objectify' social science. For us, such an attempt to transcend reflexivity in the pursuit of more 'objective' knowledge is doomed to failure (Kim, 2010). This is, in our view, because the pervasiveness of reflexivity renders such a quest fruitless. The sciences—including the social sciences—have to live *with* essential reflexivity, not attempt to eliminate it in the search for a perfect form of observation, measurement and description.

While all aspects of reflexivity are epistemic, that broad designation needs to be addressed in a more detailed way. There are several major strands in research reflexivity that need to be identified and exemplified. Here, therefore, we discuss the following sources of reflexivity: disciplinary reflexivity; methodological reflexivity; reflexivity of membership; textual reflexivity; positional reflexivity. These are all predicated on the mutual implications of *what* one studies and *how* one studies it. We consider these in turn before we discuss some of the more common uses of the term, which—we argue—are more appropriately thought of in terms of *reflection* and *reflective practice*. Even here, there is ample opportunity for confusion: Alvesson and Sköldberg (2018) begin their overview of 'reflexive methodology', by using reflexive and reflective interchangeably (p. 10 ff.), even though they modify that conflation later in their book.

Disciplinary Reflexivity

Irrespective of whether social scientists identify themselves closely with one discipline, or whether they embrace a more fluidly interdisciplinary stance, the intellectual traditions that partly constitute disciplinary knowledge bear directly on the nature of research and the phenomena that it takes as its proper subject-matter. In other words, disciplines set the possibilities of research. They help to define what is worth studying, what counts as worthwhile or newsworthy subject-matter, what is worth taking seriously, and how to identify it. Such disciplinary reflexivity is simultaneously productive and constraining. It is productive, in that it can suggest fruitful lines of inquiry, and furnish the means to pursue them. It can provide the researcher with templates and exemplars that can guide the researcher—the novice in particular. At the same time, it can constrain research, precisely because those lines of inquiry can implicitly exclude, downgrade or marginalise other phenomena, rendering them 'unthinkable'.

To that extent, the reflexivity of disciplinary knowledge and tradition can be likened to the notion of a *collective thought-style* (Fleck, 1937 [1979]) or a scientific *paradigm* (Kuhn, 1996). The paradigm is not just a preference for one theory over another at a given point in time. It is a package of key ideas, key works, accepted methods, leading figures and role-models, and classic studies. A paradigm frames simultaneously what to study, the appropriate methods to identify the most relevant phenomena, and what the expected outcomes should look like. At the same time, such epistemic frameworks *exclude* potential phenomena. Recent scholarly attention to 'ignorance' and the emergent sub-field of ignorance studies drives home that point. Ignorance is not just a matter of random lacunae or temporary misunderstandings, nor does it refer to incomplete research where the results are not yet known. Rather, ignorance in this sense can be just as systematic as any body of knowledge, and the study of ignorance can be a significant contribution to our collective grasp of disciplinary reflexivity. That does not mean that any given discipline (such as anthropology, geography or sociology) will display one homogeneous array of guiding ideas. The outcomes of disciplinary reflexivity are not immutable either. Indeed, if we are to understand the reflexive framing of research, then we need to be sensitive to the internal differentiations of the discipline.

When it comes to ethnographic fieldwork, there are clear distinctions to be drawn. Anthropological traditions can frame what will 'count' as a *field* that is appropriate for fieldwork. The fields of anthropological fieldwork are not pre-given entities. They are the outcomes of disciplinary predispositions, biographical circumstances, and chance. But the disciplinary reflexivity should not be discounted. The question is always: What is, or should be, the field of your fieldwork? And the answer will vary across time and across national boundaries. To take one obvious example: British social anthropology and American cultural anthropology have historically taken divergent approaches in defining what counts as the proper object of scrutiny, and thus what counts *in* the field and *as* the field. In turn those two English-language approaches contrast with the French traditions. And so on. We do not need to rehearse all of the variations that can be identified.

One pertinent example is the development of anthropology in the Netherlands. Geographically, and like most anthropological schools, it followed the geography of colonialism, and was therefore closely aligned to orientalist scholarship, with the Dutch colonies—such as Indonesia—as the sites of field research. While it is not widely known beyond a small number of specialists, the Dutch developed a distinctive, national style of structuralism. The latter was centred particularly on the department at Leiden (Locher, 1988) and owed much to the ethnology of Marcel Mauss (Josselin de Jong, 1972). Their 'fields' and the phenomena they identified were not completely idiosyncratic, of course, but Dutch anthropology developed a distinctive canon of ethnographies.

If we then reconsider just British anthropology, it is relatively easy to see what has in the past been deemed as a worthy 'field': somewhere remote, based on settlements small enough to study at firsthand. The fields of fieldwork, for the most part, followed the geographical spread of Empire: Africa, South and Southeast Asia, Papua New Guinea, Melanesia and the Pacific. Given the preference for the remote and self-contained, islands were over-represented in classic British anthropology: Kiriwina (Trobriands), Dobu, Tikopia, the Andamans, Anafi, as were isolated rural hamlets. In recent decades, what counts as a field has changed. It is now possible for a British anthropologist to study in a field 'at home' Fields close at hand have taken their place alongside the far away. At the same time, 'the field' is much less likely to be thought of in terms of a complete account of a single social system. As well as conducting fieldwork at home, anthropologists can now study processes and institutions quite different from the phenomena studied by the discipline's founders. Equally, 'classical' anthropological ethnography defined the 'proper' topics for inquiry. Disciplinary training, and the textual arrangement of the ethnographic monograph, defined the topics of scrutiny: family, marriage and kinship; land tenure and residence; economic and ritual exchange; religion and belief systems; politics and high office. Such major themes were sufficient to capture a 'holistic' analysis of the entire social system. Moreover, local, departmental lineages shaped how those kinds of issues might be addressed. Among the major 'schools' of British anthropology, there were clear differences in subject-matter and approach as between, say Cambridge, Oxford and Manchester. Each local tradition defined its own constellations of problematics, topics and preferred analytic approaches.

As Barth et al. (2005) make clear, there are significant national characteristics that distinguish British, German, French and American anthropology. They had different intellectual roots and foundations that influenced their development for at least the first half of the twentieth century, and beyond. The increasing globalisation of academic publishing and geographical mobility may have softened some of those differences, but they are still discernible. American anthropology's course was partly set by an emphasis on 'culture', in opposition to explanations based on 'race', while British anthropology had a much greater emphasis on social structure and institutions. It went on to have a much stronger emphasis on kinship and descent, while the American tradition was reflected in 'interpretive' anthropology, sponsored by Geertz and those influenced by him (Geertz, 1973). That in turn fed into a 'textual' perspective on anthropological interpretation. French anthropology, by further contrast, inherited the mantle of Durkheim and his nephew Mauss, with a marked interest in 'collective representations', such as religious and aesthetic conventions. Structuralism, promoted by Lévi-Strauss, was a distinctive element in the French traditions. While a small number of British anthropologists assimilated French structuralism, for the most part it stood in contrast to British structural-functionalism. The roots of German anthropology lie partly with a Romantic view of 'peoples', and the development of the discipline was inevitably strongly shaped by the Nazi period. Now the point here is *not* to imply that such national traditions are hermetically sealed off from one another. Clearly they are not. And there is room for plenty of variation within such traditions (as we shall mention below). Recent and contemporary versions of anthropology have been more permeable, open to influences from multiple geographic and intellectual sources. We draw attention to national tendencies within disciplinary fields in order to emphasise that traditions, of various sorts, can shape—in various ways—what is worth studying, and how to set about it.

National traditions in qualitative sociological (and cognate) research are less well documented than in anthropology. There are, however, European research lineages in life-history and biographical research that—while poorly known in the English-speaking world—are of considerable significance. Atkinson et al. (2011) discuss the extent to which biographical research reflects the social and intellectual history of the nations in question. They discuss in particular the Italian and Polish traditions and their major protagonists. In common with yet other European research lineages, their subject-matter and approach reflect the nations' political and cultural histories. Life-histories and biographical research often address actors' narratives of disruption, dislocation and survival, marked by periods of fascist or communist rule, war and resistance (see also Bertaux et al., 2004 for narratives of survival in Soviet Russia.).

In the context of sociological ethnography, there are clear disciplinary traditions that have framed the subject-matter of ethnographic research (Atkinson & Delamont, 2005). Urban ethnography defined its subject-matter, from the origins in Chicago sociology, in terms of local—often ethnic—organisations and sub-cultures; crime and deviance; marginalisation. To those were added ethnographic studies of organisations such as schools, hospitals and workplaces. They in turn could be framed in terms of careers and moral careers; rules and rule infractions; coping strategies and routines; workers' and inmates' resistance. The sociological and anthropological traditions converged in the sociological study of 'communities': geographically defined locales that were described in terms of localism, close interpersonal ties, and traditional occupations (such as mining, fishing or agriculture).

In recent years, anthropological and sociological inquiries have become characterised by different arrays of interest. It is not our purpose to review them all. We note that the shifts in disciplinary culture have framed newer and different topics of ethnographic focus. The influences of feminism, postcolonialism, critical race theory and similar movements have redirected ethnographic attention towards issues of identity, sexuality, stigma. A pervasive if unacknowledged influence of interactionist ideas (Atkinson & Housley, 2003) has reinforced a contemporary emphasis on 'micro' phenomena and interpersonal encounters.

Self-evidently disciplinary reflexivity reflects changing research foci in the parent discipline. For instance, feminist and queer standpoints reflexively re-frame the proper subject-matter of ethnographic fieldwork. So too does the emergence of critical race theory. There is, therefore, a dialectical relationship whereby observed phenomena in the field and the key ideas that inform the discipline interact with one another. They mutually constitute what is describable, analysable, explicable, and—most generally—what counts as thinkable. In recent decades, Indigenous research and Indigenous methods have achieved considerable prominence. As well as promoting the interests of a relatively small number of Indigenous or First Nation Peoples, such a research strategy creates new objects of research, or at least aims to, by framing its aims in terms of local knowledge systems and local means of understanding.

As Bourdieu pertinently observes, the intellectual field not only provides paradigm-like frames for the design and conduct of research, it also furnishes the critical audience for that research. The researcher's 'peers' therefore evaluate the worth of the research against the collectively shared—often implicit—criteria that inform the choice and identification of researchable phenomena in the first place. The intellectual field sets the rules of the game. To that extent, therefore, disciplinary reflexivity is also a matter of power, since the capacity to define what is 'thinkable' and what might count as the right way to produce valid knowledge inscribes the power of individuals and networks to define the fields of inquiry. Such disciplinary power often resides with what Collins (1981) referred to as 'core sets'. Even in globalised networks of scholars, many core sets are comprised of relatively few influential figures, who can—quite legitimately—exercise considerable influence, as authors, editors, reviewers, keynote speakers, supervisors, mentors and research-group leaders. Shaping the intellectual climate of academic departments, journals, academies, and grant-awarding organisations is the work of core-set members. One does not need to attribute base motives to them in order to recognise the extent to which intellectual fields can be collectively moulded through the dispositions of such influentials. Their intellectual influence is both productive and limiting. They can promote particular lines of inquiry and styles of investigation, but they can also limit access to alternative, heterodox approaches.

Fuller (2016) enunciates a thorough critique of academic disciplines, suggesting that influential members become intellectual rentiers, exacting 'tribute' in the form of adherence to formulaic versions of knowledge production and the appropriate citations. (The metaphor of 'paying one's dues' captures this perceived obligation.) In Fuller's terms, a paradigmatic text, such as a methodological prescription or an exemplary monograph.

> ...effectively 'sublimates' the details of specifically situated activities into an abstractly worded version that may be easily imported by others for use and development. This appropriation is subject to a payment of tribute—another form of rent—the giving of credit, on which the academic citation culture is based. This is the currency in which careers are nowadays made or broken. (Fuller, 2016: 41)

Fuller goes on to suggest the irony that university teachers want their students to 'decode' academic texts just so that they can replicate that coding in their own work. This redoubles the payment of academic credit and—more importantly for our argument—amplifies the effect of disciplinary reflexivity, shoring up still further what is and is not thinkable and how knowledge is justified. Fuller recapitulates some anthropological observations on patron-client relationships, and the parallels with academic patronage, that create the conditions for disciplinary boundaries and genealogies (e.g. Boissevain, 1974).

Such constraints depend on strongly classified and framed fields of research and publishing. Those terms derive from Basil Bernstein's sociological analysis of curriculum and pedagogy (Atkinson, 1985). Classification refers to the principles of inclusion and exclusion in a given intellectual field. When classification is strong, then there are powerful symbolic and institutional boundaries surrounding that field. Framing refers to the relative strength of pedagogy and mechanisms of socialisation. The mechanisms that Fuller identifies imply strongly framed and strongly classified disciplinary fields. In such symbolic systems, personal and intellectual loyalty is fostered. It sustains that sense of academic tribalism and territorialism identified by Becher and Trowler (2001). The reflexive constitution of disciplinary loyalties is sustained in academic fields through the strongly framed pedagogy of postgraduate 'training', not least in methodological training: 'Through the stability of pedagogical practice and pedagogical knowledge, taken-for-granted forms and contents of scientific thought are transmitted from generation to generation' (Delamont et al., 2000: 13). Enculturation into disciplinary knowledge promotes close identification with leading figures, local loyalties and intellectual identities based on lineages and genealogies. The practice of peer review can also buttress that sense of genealogical authority, where 'a common condition for publication is that the author must add references to other putatively related work that the author may not have read—let alone been influenced by—but serves to reinforce the peers' sense of the lines of epistemic descent' (Fuller, 2016: 63–64).

Our identification of disciplinary reflexivity does not imply stasis. Clearly, disciplinary contents and methods do change over time. New trends and perspectives emerge, promoted by key actors and sustained through institutionalised means—textbooks and edited collections, conferences and journals. We need to think of disciplinary cultures in terms of mutating cultures of knowledge, with localised mechanisms of classification and framing. Those cultures of knowledge-production are supported through the reflexivity of method, to which we turn in the next chapter.

The collective examination of methodology and standpoint parallels the recurrent interest in what is sometimes called *methodography*. That is, a close examination of how the methods of the social sciences are deployed, and how they shape their objects. In many ways that is the overall tenor of Cicourel's landmark contribution, *Method and Measurement* (Cicourel, 1964), the general message of which was that sociologists were far too reliant on measurements (not just quantitative) that imposed unexamined assumptions in categorising phenomena. The argument can be applied to all methods. Methodography satisfies—in part—Bourdieu's injunction that academics should include their own practices within their objectifying scrutiny. A recent example is the critique of 'coding' texts in cultural analysis by Biernacki (2012). Methodography itself needs to be examined more closely and more extensively, if only to prevent the practice of examining reflexivity unduly individualistic.

References

Alvesson, M., & Sköldberg, K. (2018). *Reflexive methodology: New vistas for qualitative research* (3rd ed.). Sage.

Atkinson, P. (1985). *Language, structure and reproduction: An introduction to the sociology of Basil Bernstein*. Methuen.

Atkinson, P., & Delamont, S. (2005). Qualitative research traditions. In C. Calhoun, C. Rojek, & B. Turner (Eds.), *Handbook of sociology* (pp. 40–60). Sage.

Atkinson, P., & Housley, W. (2003). *Interactionism*. Sage.

Atkinson, P., Seweryn, A., & Tirini, S. (2011). Knowing selves: Biographical research and European traditions. *International Review of Qualitative Research, 4*(4), 461–488.

Barth, F., Gingrich, A., Parkin, R., & Silverman, S. (2005). *One discipline, four ways: British, German, French, and American anthropology*. University of Chicago Press.
Becher, T., & Trowler, P. R. (2001). *Academic tribes and territories* (2nd ed.). Open University Press.
Bertaux, D., Rotkirch, A., & Thompson, P. (Eds.). (2004). *On living through Soviet Russia*. Routledge.
Biernacki, R. (2012). *Reinventing evidence in social inquiry: decoding facts and variables*. Palgrave Macmillan.
Boissevain, J. (1974). *Friends of friends*. Basil Blackwell.
Bourdieu, P. (1990a). *In other words: Essays towards a reflexive sociology*. Stanford University Press.
Bourdieu, P. (1990b). *The logic of practice*. Polity Press.
Bourdieu, P., & Wacquant, L. J. D. (1992). *An invitation to reflexive sociology*. Polity Press.
Cicourel, A. V. (1964). *Method and measurement in sociology*. The Free Press.
Cohn, N. (1970 [1957]). *The pursuit of the millenium*. Paladin.
Collins, H. M. (1981). The place of the 'core set' in modern science: Social contingency with methodological propriety in science. *History of Science, 19*(1), 6–19.
Delamont, S., Atkinson, P., & Parry, O. (2000). *The doctoral experience: Success and failure in graduate school*. Falmer.
Elias, N. (1970). *What is sociology?* Columbia University Press.
Fleck, L. (1937 [1979]). *The genesis and development of a scientific fact*. University of Chicago Press.
Fuller, S. (2016). *The academic Caesar: University leadership is hard*. Sage.
Garland, D. (2014). What is a 'history of the present'? On Foucault's genealogies and their critical preconditions. *Punishment and Society, 16*(4), 365–384.
Geertz, C. (1973). *The interpretation of cultures*. Basic Books.
Hess, D. F. J. (2011). Bourdieu and science and technology studies: Toward a reflexive sociology. *Minerva, 49*, 333–348.
Josselin de Jong, P.E. de (1972). Marcel Mauss et les origines de l'anthropologie structurale hollandaise. *L'Homme, 12*(4), 62–84.
Kim, K.-M. (2010). How objective is Bourdieu's participant objectivation? *Qualitative Inquiry, 16*(9), 747–756.
Koselleck, R. (1998). [1959]). *Critique and crisis*. Berg.
Kuhn, T. S. (1996). *The structure of scientific revolutions* (3rd ed.). University of Chicago Press.

Locher, G.W. (1988). J.P.B. de Josselin de Jong en het Leidse structuralisme. In P. Hovens & L. F. Triebels (Eds) Historische ontwikkelingenin de Nederlandse antropologie. *Antropologische Verkenningen, 7*(1–2), 51–74.

Maton, K. (2003). Reflexivity, relationism, and research: Pierre Bourdieu and the epistemic conditions of social scientific knowledge. *Space and Culture., 6*(1), 52–65.

May, T., & Perry, B. (2017). *Reflexivity: The essential guide*. Sage.

Mendel, I. (2006). Mannheim's free-floating intelligentsia: The role of closeness and distance in the analysis of society. *Studies in Social and Political Thought, 12*, 30–52.

Smith, D. (2000). *Norbert Elias and modern social theory*. Sage.

Smith, R.J. and Atkinson, P. (2016). Cicourel's *Method and Measurement in Sociology*, fifty years on (with responses by Aaron Cicourel and others). *International Journal of Social Research Methodology, 19*(1), 99–110.

Swartz, D. (1997). *Culture and power*. University of Chicago Press.

Szakolczai, A. (2000). *Reflexive historical sociology*. Routledge.

Vandenberghe, F. (1999). "The real is relational": An epistemological analysis of Pierre Bourdieu's generative structuralism. *Sociological Theory, 17*(1), 32–67.

Woolgar, S., & Ashmore, M. (1988). The next step: An introduction to the reflexive project. In S. Woolgar (Ed.), *Knowledge and reflexivity: New frontiers in the sociology of knowledge* (pp. 1–11). Sage.

Wacquant, L. J. D. (1989). Towards a reflexive sociology: A workshop with Pierre Bourdieu. *Sociological Theory*, 7(1), 26–63.

3
Methodological Reflexivity

Abstract Epistemic reflexivity means that the methods used to describe, classify and measure phenomena contribute to the construction of those phenomena themselves. The chapter, focused primarily on qualitative methods, examines some key aspects of methodological reflexivity: ethnographic fieldwork, interviewing, coding and transcribing among them. The chapter then describes the reflexivity of classifications and categories, and discusses researchers' positionality in relation to reflexivity. The significance of textual representations is also discussed.

Keywords Methodological reflexivity · Ethnography · Interviewing · Coding · Transcription · Positionality · Textual reflexivity

Since reflexivity refers to the inescapably reactive nature of any research intervention, it is obvious that the research methods used will reflexively shape the kinds of phenomena that are identified, classified and measured. The topic is a rich and complex one. But given the nature of reflexivity, methodological reflexivity is perhaps the most central of topics, and is one that is immediately within the grasp of all active social

E. M. Whitaker and P. Atkinson, *Reflexivity in Social Research*,
https://doi.org/10.1007/978-3-030-84095-2_3

researchers. Methodological reflexivity applies with all methods—standardor innovative, quantitative or qualitative. It is by no means necessary for the assumptions of methods to be 'hidden' and therefore needing to be uncovered. Many of the assumptions of method are overt, but that does not exempt them from reflexive framing of the phenomena. Equally, it probably needs reaffirming here, a consideration of reflexivity does not imply the search for a perfect, neutral and transparent method. There exists no such method.

Our discussion will be focused on varieties of 'qualitative' research methods, as reflexivity is most closely associated with such methods. This does not mean that we endorse all versions of so-called 'qualitative'. It certainly does not mean that we dismiss all 'quantitative' research strategies in the social sciences either. But it is important to examine methodological reflexivity in some depth here. It is too readily assumed in some quarters that qualitative work is inherently superior by virtue of distinctive qualities, of which reflexivity is a key feature. It is also important for us to examine some of the methodological assumptions and implications of various research strategies and traditions. We need to highlight the fact that there is no research approach that is 'pure' in giving us access to phenomena without the mediation of 'apparatus' and presuppositions. Appeals to 'naturalism' do not solve that, nor do rejections of any and all conventions of research strategy and method.

We discuss a selected number of research strategies: we make no attempt to list and examine each and every method or research strategy. The crucial point here is that 'methods' create 'data'. Now some extreme positions in qualitative social research may suggest the 'death of data' (Denzin, 2013) and try to expunge all reference to such 'traditional' research concerns as 'evidence' or 'rigour', embraced in several 'post' positions (see for instance St. Pierre, 2011, 2015). Here, our use of 'data' is intended to imply no particular orientation towards epistemological disputes. Rather, we use the term to mean any organised, collated array of material that is susceptible to classification and categorisation in the interests of deriving research-based knowledge. There is no need to dismiss all such material, or interest in it, as 'positivist'. More generically still, methods create the possibility of research, framing *what* is to be studied, *how* it is to be studied, *what* will count as a plausible outcome of the research, and *how* it will be evaluated. Issues of method are, therefore, productive in the creation of knowledge, as well as implicitly constraining what knowledge is thinkable.

We do not trudge through every single 'method' here, but we discuss a selected number of them by way of illustrating the general features of methodological reflexivity. In doing so, we emphasise once more that reflexivity here is to be understood as the essential relationship between methods and the phenomena they describe; it is not a property of some methods rather than others, nor is it a matter of 'reflecting' (however critically) on those methods and their applications. We discuss participant observation, interviews, coding, transcription in particular, and then discuss the apparatus of social research more broadly again.

Participant observation and fieldwork. The core of ethnographic research, participant observation is perhaps the most self-evidently reactive, depending as it does on the direct engagement of the researcher with their research hosts. Our discussion does not rest on the glib suggestion that participant observation is inherently flawed because actors 'behave differently' in the field under study. This is not a 'simple' matter of the researcher's presence in the field. It rests on the fact that the phenomena that are describable, and the events that are reportable, are largely the outcomes of the encounters that are possible and achievable in the field. Indeed, the field is the outcome of successive negotiations between the ethnographer and groups or individuals. At a fundamental level, what counts as the 'field' of fieldwork is not a given. Fields of research are not naturally occurring entities. Obviously, there are some institutions that are relatively self-contained—or apparently so—but in practice fieldwork does not encompass the entire hospital, prison, school and so on. Moreover, the actual fields of practical research are constructed through the kinds of concrete engagements that the ethnographer can actually engage with in the course of weeks or months 'in the field'. Among anthropologists, the construction of the field has become a topic of some prominence (see Amit, 2000), not least because contemporary anthropologists increasingly work on what are—for that discipline—non-traditional sites. As anthropology has increasingly turned towards fieldwork 'at home', so they have been led to reflect more self-consciously on 'fields'. But that is not the most fundamental issue. Ethnographers, however 'naturalistic' their research approach, cannot claim exemption from methodological reflexivity. In concrete terms, their 'fields' of fieldwork are practical accomplishments on the part of the ethnographer and through processes of negotiation with their research hosts.

We know, from ethnographers' autobiographical accounts, and from the methodological literature, that one cannot simply negotiate 'access' to a chosen social setting on the basis of a single once-and-for-all transaction. The pursuit of ethnography nearly always demands a continuous process of negotiation and re-negotiation. 'Access' and relationships in the field require interactive processes of work. Even if such interpersonal work is not explicitly framed as 'access negotiations', ethnographers and their hosts are often engaged in repeated negotiations that implicitly define borders: settings that are open to observation and others that are not; matters that can be talked about and those that cannot; actors who are willing to be observed and engaged with and those who are reticent or who refuse. These interpersonal dynamics partly define the contours of what 'the field' is, and what action can be participated in and observed. The traditional vocabulary of 'access negotiations' itself implies a territorial boundedness, a kind of checkpoint mode of ingress. It conjures up the 'field' as an already bounded entity. It does also imply that the field is a collective entity. The contemporary replacement of 'access' with the 'informed consent' required by contemporary ethics committees implicitly constructs the field differently. Rather than the shared, collective entity, the usage constructs 'the field' as a series of individualised actors, enrolled as a more-or-less unrelated 'participants' and divorced from their collective engagements (Atkinson, 2009).

As we have just said, what 'the field' is *in practice* rests on the ethnographer's pragmatic engagements with others. It is not an inert arena or background. What we call the field of fieldwork is in itself a complex assemblage of networks and relationships. It is physically and symbolically partitioned. It is divided into domains that are more or less 'private' or 'public', 'backstage' or 'frontstage'. Some aspects of a field are the specialist domain of selected groups or individuals. They are segmented according to divisions of labour. The early work of the Ardeners on 'muted groups' was a salutary reminder of the reflexive construction of fields (Ardener, 1972, 1975). It emphasised the extent to which there could be multiple cultural realities in play within 'the same' setting. The 'muted' group—in the original case, women—do not merely inhabit 'their own' cultural and social space but adapt to that of the dominant (male) group (Delamont, 1989). This is analogous to the idea of 'double consciousness' pioneered by W. E. B. Dubois, who used the term to capture key aspects of Black Americans' experiences of living in a racialised society (Bruce, 1992). And for our purpose, it reminds us that the ethnographer will 'discover' a version of the local culture depending on what segment(s) of the field she or he is able to negotiate adequate access to.

In recent decades, much more of ethnographic fieldwork has been conceptualised in terms of 'collaborative' relations, and 'the fields' of fieldwork conceptualised accordingly. The collection of essays edited by Estalella and Criado (2018), for instance, are derived from projects based specifically on 'collaboration' on the part of ethnographers with members of professional and epistemic communities. This is a form of fieldworking that not only changes the social relationship in the field, but also transforms how those fields are conceptualised: 'In these situations, the traditional tropes of fieldwork encounter (i.e. immersion and distance) give way to a narrative register of experimentation, where the aesthetics of collaboration in the production of knowledge substitutes or intermingles with the traditional trope of participant observation' (Criado & Estallella, 2018: 2). This is not just a different *purpose* for fieldwork, but a re-evaluation of what the field and what fieldwork actually consist of. Those sites of collaborative knowledge-production have been described as *para-sites*, while those collaborators who are themselves engaged in knowledge formation and exploratory practice, have been described as doing 'para-ethnography'. Such skilled collaborators can include designers, film-makers and video-artists, musicians, technicians and scientists. The nature of collaboration implies a radical re-evaluation of the 'field'. As Pink says, in a concluding overview of that edited collection:

> What was conventionally called 'the ethnographic field' is ongoingly made and remade through our active participation as ethnographers in collaboration with research participants, other stakeholders in research and future

> readers and viewers. It is clear that the ideas of ethnography being something that is fixed in (even multiple or moving) field sites, are insufficient. Instead the temporalities and sites of the emergence of ethnographic knowing stretch beyond this; various collaborators may shape it—they are part of what I have elsewhere called 'the ethnographic place…' (Pink, 2018: 201–202)

Those ethnographic places are far from arbitrary, but equally, they are shaped and framed by the everyday work and engagements of ethnographers and participants alike. The fields of fieldwork are, therefore, not given 'naturally' nor do they pre-exist the work of the ethnographer, whose practical and intellectual work transmutes the multiple and contingent events of everyday life into some semblance of a coherent assemblage.

The nature of ethnographic fieldwork adds layers of 'reflexivity' to the overall anthropological or sociological project. They reflect the self-evident fact that the 'subjects' of research—the actors with whom one interacts in the field—are themselves agents who enact their own interpretations and performances of their own social world. The reflexive nature of observation and participation rests on the extent to which social actors are themselves active in constructing their own versions of social reality. Ethnographic observations are not performed on otherwise inert objects. Social and cultural phenomena are produced through actors' everyday work of action and interpretation. Consequently, we shall need to pay attention to a principle of reflexivity that reflects the intensely social and dialogic foundations of ethnographic work. As a number of authors have suggested, performances such as rituals and other collective observances may be as much reflexive constructions of the field and of the culture as the work of the ethnographer herself (Myerhoff & Ruby, 1982). There is, therefore, a double process at work—as the ethnographer reflexively constructs cultural categories that are simultaneously being constructed by the participants themselves. Geertz's interpretation of the Balinese cockfight is often invoked in this context. Such reflexive enactments are then, of course, available for further textual or performative reconstructions (Schechner, 1982; Turner, 1982).

This therefore leads us to a related, but different array of uses of reflexivity, from within the field of performance and language studies. In its most general sense, here it is used to refer to actors' *reflexive consciousness*. Babcock (1980) summarises a number of usages in that context. First, it can refer to the capacity of language to refer to itself, in the form of metalanguage. Secondly, it refers to the capacity of actors to be aware of themselves as subjects and as objects of others' awareness. Third, it can refer to the capacity of participants in events like rituals to use the resources of culture to comment on or even to transform that culture itself (see Bauman, 1986, 1989; Berger & Del Negro, 2002). We shall comment on general ideas of reflexive consciousness later, at this point noting that it relates these perspectives on performance with broader sociological themes, not least those of symbolic interactionism (Atkinson & Housley, 2003), derived from the social psychology of George Herbert Mead (1934), that affirms that the capacity for self-awareness as a defining trait of human being and the basis for social interaction. This is a generic perspective by no means confined to the pragmatist tradition of symbolic interactionism, however. Writing exclusively about anthropology, for instance, Evens et al. (2016) suggest that reflexivity is 'a defining feature of the being and becoming of the human' (p. 1) that they also link directly to actors' capacity to experience both the self and the other. And in an equally generic sense—also from anthropology—Myerhoff and Ruby (1982) assert: 'Reflexive, as we use it, describes the capacity of any system of signification to turn back upon itself, to make itself its own object by referring to itself: subject and object fuse' (p. 2). This latter formulation is close to our generic understanding of reflexivity: the inevitable fact that our 'apparatus' of research simultaneously justifies the research act, and generates the phenomena that it describes.

Interviews and narratives. The interview is an extraordinarily common feature of contemporary social research, especially in what currently passes for 'qualitative' work. Now the interview is not a single entity, and there are many forms of interview. More fundamentally, interviewing presupposes particular relationships between the researcher and the interviewee. Moreover, the interview can itself inscribe particular expectations or constructions of what kind of 'data' will be forthcoming.

Such presuppositions can also project kinds of social actors—speaking subjects who will display selves and identities in an appropriate manner. As Davies (2008) points out, the interview is not a culturally universal phenomenon. Moreover, even within 'Western' culture, the interview is a highly variable encounter. It can range from a simply 'vox pop' interview with the man-or-woman-in-the-street, to feature-length interviews conducted by writers with celebrities and public intellectuals. Atkinson and Silverman (1997) suggested that the widespread reliance on interviewing in 'qualitative' research is a symptom of what they called the 'interview society'. The culture of the interview, they suggested, is preoccupied by a search for how respondents 'feel' about their personal experiences. The interview is thus constructed as a means for the exploration and revelation of an otherwise private, inner self. In the same way 'qualitative' interviewing can be aimed at uncovering the interiority of a romanticised subject. And as Whitaker and Atkinson have argued subsequently (2019), the interview society and the culture of interviewing should now also be thought of in terms of therapeutic encounters. The act of enunciating one's experiences, emotions and reactions is confessional, and—like confessions—such interviews can be constructed as liberating for the interviewee. While there are clearly differences in emphasis, such qualitative, extended interviewing often yields 'narratives' of personal experience.

The collection and interpretation of narratives has become a pervasive research commitment among qualitative researchers. It can be justified in terms of 'giving voice' to research participants who can be read expressing themselves 'in their own words'. The search for narratives and narrative significance includes a number of key presuppositions. Most importantly, it can assume a particular kind of speaking subject and a distinctive array of spoken performances. It also reflects the kinds of implicit assumptions about 'the interview society' that Atkinson and Silverman (1997) identified. They suggested that in their wholesale adoption of the personal, experiential interview, many sociologists were complicit in reproducing the cultural tropes of an interview society obsessed with personal confession and biographical self-exposure. It is predicated on a particular kind of speaking subject—one who can and will articulate 'experiences', 'feelings' and 'personal' aspects of their lives. The research is often based on the assertion that 'narrative' is a fundamental method whereby human actors order their lives and their experiences. Narrative becomes a master method: simultaneously the method of ordinary lives, and the method for investigating such lives. The ideal, imagined narrator is one who is capable of generating extended accounts of personal development, epiphanies and turning-points, descriptions of persons and events. Now we are not dismissing 'narratives' out of hand here. Clearly, narrating and telling stories are social actions, speech events, in their own right and deserve to be studied appropriately. But we are concerned here with how a commitment to narrative research reflexively shapes the researcher's search for distinctive kinds of narrators and narratives.

The philosopher Galen Strawson has made a significant intervention in the now vast literature on narrative and identity (Strawson, 2004). He was responding to what he saw as a powerful and pervasive emphasis on narrative in social and psychological disciplines. He identifies what he calls the 'psychological Narrativity thesis' (the capitalisation is Strawson's). This can be traced in a number of authors and disciplinary fields. Its fundamental thesis is, however, constant across them: humans are inherently narrative beings. This is, Strawson suggests, often linked to a normative thesis: that a narrative understanding or construction of one's life is a good thing. Now there is nothing to suggest that actors do not produce narratives and stories—clearly they are common forms of spoken performance. The reflexive issue rather resides in the implication that, as Strawson notes, the narrative approach often implies a particular kind of speaker, and through the relevant analytic approaches, uncovers such speakers. This is a mode of thought and analysis that owes much to the culture of confession. As Gubrium and Holstein (2003) summarise it: 'The interview society not only reflexively constructs a compatible subject, but fully rounds this out ontologically by taking us to the proverbial heart of the subject in question' (p. 29). And they go on by expanding on that idea:

> This reveals the romantic impulse behind the interview and the interview society. If we desire to 'really know' the individual subject, then somehow we must provide a means to hear his or her genuine voice. Superficial discussion does not seem to be adequate. Many interviewers

> explore the emotional enclaves of the self by way of 'open-ended' or 'in-depth' interviewing. Although, technically 'open-endedness' is merely a way to structure the interview process, Atkinson and Silverman suggest that the term also flags a particular social understanding, namely, that the true, internal voice of the subject comes through only when it is not externally screened or otherwise communicatively constrained. (ibid.)

In other words, the implication is that actors ought to produce narratives. Moreover, this in turn seems often to imply that the narratives they produce ought to have a certain kind of quality. In practice, this seems to imply narrative fluency and elaboration. The successful narration is produced by a narrator who is capable of appearing as a protagonist in an extended account or series of accounts, in which events are not merely recounted or chronicled. They are subject to narrative elaboration. They are reflected upon and evaluated.

It is clear that interviews can never be regarded as a purely neutral means of data-collection. The interview projects particular kinds of 'information' and—more fundamentally perhaps—particular kinds of respondent. The outcome of the interview (inner feelings or cultural categories, say) is in turn predicated on the researcher's intention and the form of the questioning. The interview society and the use of the extended personal interview implicitly expects the interviewee to articulate her or his personal, even private, responses. The production of personal stories or narratives is also a preferred outcome for some uses of the extended or life-history interview. Again, the idea of the life-history, and the construction of the interview so as to elicit it, implies a particular kind of personal development, a moral career for the respondent, and also what counts as research evidence. This is not to say that such expectations and projections are always 'wrong'. (There is room for argument over which model of culture, or social actors is more appropriate.) But it *is* a reflection of the extent to which a particular ethnographic or, more generally qualitative style of interviewing can reflexively construct not just 'data' but also distinctive kinds of speaking subject. Hence, this is a key exemplar of methodological reflexivity, shaping not just information but projecting the kinds of social actors or subjects implicitly preferred by the method.

Coding. The 'coding' of qualitative data is a pervasive analytic strategy in contemporary social research. It is a form of measurement, in that it explicitly calls for the categorisation of data and the allocation of segments of research materials into equivalence-classes. Here—as we have already established—'measurement' is not specifically tied to procedures of quantification, although the coding of things like interview transcripts *can* be used to develop frequency counts and the like (transforming intuitions concerning typicality, routine-ness, or rarity into scores). Varieties of coding are fundamental to the development and implementation of varieties of 'grounded theory' (Charmaz, 2014), although the underlying logic of abductive logic is *not* dependent on such coding procedures. That is to say that virtually all methods of creating and managing data for such analytic approaches depend upon at least implicit 'coding', to the extent that the analyst searches her data for 'instances', 'examples', or even 'quotes' that stand for more general categories of phenomena. Indeed, the decisions and acts of pulling together segments and examples, *creates* those very categories.

Saldaña (2016) provides a valuably extended account of coding for qualitative researchers, and is explicit about the processes of decision-making that go into coding as an analytic process: 'To codify is to arrange things in a systematic order, to make something part of a system or classification, to categorize. When you apply and reapply codes to qualitative data, you are codifying—a process that permits data to be divided, grouped, reorganized and linked in order to consolidate meaning and develop explanation ....' (p. 9). Further decision-making groups codes into 'categories' that inscribe the 'meaning' that brings together that collection of segments. It is clear enough from Saldaña's treatment that the derivation of codes and categories necessarily reflects the analyst's theoretical and substantive interests. Indeed, he describes a series of procedures whereby instances, codes and categories are progressively synthesized in generating a relatively large number of codes and a relatively small number of major analytic themes (p. 25). So codes create categories, which are not natural types but are constructed through successive rounds of practical decision-making. Such classes of phenomena are sometimes referred to as *themes*, leading to varieties of *thematic analysis* (TA). Now thematic analysis is most frequently applied—like a great deal of qualitative work—to transcripts of interview talk. As its advocates insist, such 'themes' do not 'emerge' through some process of spontaneous generation: they are not inherent properties of the form and content of the talk. Hence, they are not 'found' so much as created. They reflect a process of decision-making on the part of the analyst. The construction of categories and themes depends on the judgement of the analyst(s). Indeed, invoking the reflective version of reflexivity, Braun and Clarke (2019), two leading proponents of TA make explicit its reliance on the researcher's judgements. Indeed, they celebrate a reliance on 'subjective' judgements: 'We intended our approach to TA to reflect our view of qualitative research as creative, reflexive and subjective, with researcher subjectivity understood as a resource...rather than a potential threat to knowledge production... (p.). In ways we shall return to later, we do not equate decision-making—informed by disciplinary knowledge and methodological convention—as equivalent of 'subjectivity', except in the trivial sense that it depends on the exercise of agency.

Housley and Smith (2011) are among the few scholars in recent years to have studied processes of coding. They note at the outset that the practice of coding entails three stages:

> The noticing of relevant phenomena, the collection of examples of those phenomena, and analysing those phenomena to find commonalities, differences, pattern and structure.... Coding, then, is active in the transubstantiation of raw data into sociological findings and fact. (p. 419)

Apart from the fact that we prefer a slightly different terminological usage, suggesting that procedures like transcription and coding transform raw materials into 'data', we concur: the act of coding *makes* categories and classes, whereby researchers—individually or in teams—bring their social-scientific interests to bear, identify segments of action that seem to be evidence for them, and then use the codes to 'find' the self-same phenomena. (This is an over-simplification, and makes qualitative researchers sound more naïve than they really are, but it is a reasonable summary of how many research projects are in fact conducted.)

This is not in itself a matter of critical comment. Irrespective of the terminologies, and specific procedures that are involved, some such analytic work is pervasive. Unless one is to reproduce all of the 'raw' materials (transcripts, fieldnotes, documents), then some kind of categorisation is inevitable. Even if unacknowledged, the analyst implicitly creates equivalence-classes in grouping material into categories. Indeed, at a fundamental level, this is one of the ways in which those raw materials are transmuted into *data* that can be invested with analytic significance. It is that process of transformation that is the essence of any and all methodological reflexivity. In the case of coding it reflects an underlying assumption that patterns and collocations of 'meaning' can be constructed through multiple readings and engagements with transcripts and other materials.

The work of coding is, of course, most famous because of Garfinkel's study of it (Garfinkel, 1967). He discusses the work of interpretation that informed research assistants' coding practices, based on a study of medical records. He notes that there was a strong element of circularity in that work: in order to develop and apply codes to the research materials, the coders had to rely implicitly on their prior understandings of how clinical organisations work—which was precisely what the study was intended to uncover. So the transformation of raw materials into data depended on judgements about what those data were intended to display.

Transcription. Many contemporary approaches to social research call for the transcription of recorded speech and other activities. In and of itself, transcription is not a 'method', any more than is 'coding'. But like coding, transcription is a technology for the transformation of raw materials into 'data' that are susceptible to management, analysis and reproduction. Transcription is associated with different approaches to the analysis of speech, which call for different levels of detail and sensitivity. There is no one 'correct' way to transcribe recorded talk: all methods are conventional, and their appropriateness depends on the level of delicacy called for by the researcher's analytic purposes. Conversation analysis, for example, normally demands a much more detailed level of transcription than does the representation and analysis of personal narratives or life-history accounts. That does not mean, however, that the former is 'better' than the more ordinary orthography used by ethnographers and some narrative analysts. Levels of delicacy in transcribing and the symbols used can be more or less well adapted to the analyst's purpose(s), but none is absolutely and perfectly mimetic. Indeed, the point of this discussion is—once again—not to suggest that any one technique is better than any other. Rather, we use this to exemplify our general point: that methods and techniques of data creation frame the phenomena that they describe or measure.

Hammersley (2010) makes our point when he discusses whether one can regard audio-recordings and their transcripts as 'given' data, a view that their advocates' use of the term 'naturally occurring data' might give rise to. Hammersley identifies a number of decisions that have to be made in generating transcripts. From our point of view, the following is most significant:

> How to represent recorded talk. Here, there is variation according to whether the emphasis is on capturing the actual sounds made or on identifying the words used and presenting these via traditional orthography. More specifically, there are decisions about whether to try to represent such features as intonation, pitch, amplitude, and pace of talk. Associated with this is the issue of whether to aim at capturing distinctive forms of language use, such as dialects. And, if pronunciation is to be represented, there is then the question of how to do this: whether through deviant spellings within traditional orthography or via a phonetic transcription system. (p. 556)

Hammersley's point about orthography, words, and (lack of) dialect is significant. The construction of transcripts for conversation and discourse analysis does *not* normally use the International Phonetic Alphabet—which approximates to an 'etic' system, relying on the 'emic' system of spelling and division of utterances into words. It would, as a consequence, be impossible to take a contemporary transcript and from it reconstruct—in order, say, to read aloud—even approximately how the original sounded. One would need to be told in advance what accent (rather than dialect) was being used. One would also have to be given direction as to the pitch and timbre of the voice. Moreover, the transcription is not a mechanical process. Hearing 'words' and hearing 'what was said' depends upon a degree of knowledge about the discourse, and even about the content of the talk itself. So the making of data here depends to a considerable extent on the kinds of methods (techniques of hearing) that the analyst brings to bear on the recordings. As Hammersley concludes:

> What we have in transcripts treated as data, then, are not representations of sounds recorded but rather of words heard that are components of utterances that are taken to represent actions of particular kinds: along with descriptions of relevant associated features that are designed to aid the process of inference. Both transcripts themselves, and what they are taken to show, are indeed constructed, in an important sense, but they also rely upon what is given when we listen to or watch recordings. (p. 563)

This is precisely the wider sense of methodological reflexivity: neither 'pure' construct nor 'given' data. Data always have to be constituted *as* data, and this involves some technology or technique in the process of transmutation and framing. There is nothing bogus in this, and pointing out what reflexivity means in this context is not an exercise in debunking the research.

The Apparatus of Social Research

Techniques of data collection (such as interviewing) and analysis thus shape the kinds of information and inference that are possible. They also exclude other possibilities that fall outside the possibilities that such techniques facilitate. Technologies of research can reflexively exert a strong influence on research methods, and hence of the kinds of data and analysis that they support. For instance, Uprichard et al. (2008) examine how SPSS (the statistics package) works as an inscription device. They trace the parallel development of SPSS and quantitative sociology in the United Kingdom (as it is practised and taught). They describe how the software (which is of course predicated on the digitisation of data, and the migration of the software from mainframe computers to personal computers) implicitly promotes some versions of analysis and inference, while—equally implicitly—down-playing others. In general terms, the trajectory from calculating things by hand, or with mechanical calculators, through punch cards and card-sorters, to punch cards and Fortran-based overnight data analysis, to SPSS through a PC implies a very different orientation towards data management and analytic procedures. As Uprichard, Burrows and Byrne also note, there is a similar story to be traced for qualitative sociology. Computing has taken centre stage there too. If SPSS came to be the industry standard, then a small number of commercial qualitative packages did too. Notably, Nvivo and Atlas-ti. While it would be wrong to imply technological determinism, it is equally undeniable that, as with statistical packages, Computer-Assisted Qualitative Data Analysis Software (CAQDAS) inscribes particular ways of working with data (fieldnotes, interview or focus-group transcripts, documents and images). Each of the packages implies its own distinctive way of thinking and working. Each has many functions. We are not implying here that there is anything inherently 'wrong' in basing software on a particular style of working, simply that the software itself is predicated on working methods. Of the market leaders, Nvivo was predicated in its original design on hierarchical relationships between codes, while Atlas-ti implied more heterarchical relationships. In the final analysis, there may be little difference in the outcomes of qualitative research using those and similar packages, but they do reflexively frame how 'analysis' is to be conceptualised and conducted. In turn they imply a way of approaching 'data' that differs markedly from, say, discourse-analytic approaches to similar kinds of data.

We need collectively to pay close attention to the practicalities of data analysis in order to understand how analytic decision-making reflexively constructs not just data, but the key ideas that derive from the research process. 'Analysis' is not a separate practical or intellectual process from the other aspects of the research enterprise. Analytic reflexivity extends the connotations of methodological reflexivity. It reflects the fact that the processes of analysing data are themselves constitutive of what problems, issues and phenomena will be 'discovered'. It emphasises the fact that the researcher's interpretive interactions with the data are implicative of what 'findings' will in fact be 'found'. It will influence what phenomena are singled out to be reported and treated as significant. Mauthner and Doucet (2003), for instance, suggest that this aspect of work has been under-played in discussions of reflexivity, arguing that '...reflexivity is not confined to issues of social location, theoretical perspective, emotional responses to respondents, and the need to document the research process, aspects of reflexivity which are highlighted in the current literature' (p. 418). The analytic procedures that are brought to bear and the decisions that inform them are also matters of reflexivity.

Technological developments and instrumentation impinge on qualitative research just as much as they do on quantitative methods. While ethnographers have used photography, film and sound recordings since the early years of the last century, the opportunities for data collection have been transformed in the era of digital technologies and small devices. It is now possible to collect still and moving images, record sound, record material culture, all in ways unimaginable until relatively recently. Such technical affordances in turn make possible emergent specialisms and forms of expertise. Such opportunities create newer classes of phenomena that can be captured and analysed. They can also help to create newer domains of academic specialisation, whereby sensory, spatial and material ethnographies can be brought into being, and hence novel categories of researchable phenomena.

Methodological reflexivity is clearly evident in processes and outcomes of *classification*. Cicourel's discussion of measurement, as we have acknowledged, was predicated on the view that any system of classifying and categorising is a form of measurement. Classificatory systems frame the phenomena they describe, and such systems of measurement may then arc back to determine practice and behaviour. Bowker and Star (1999, 2000) examine a variety of classificatory systems and their consequences. One of their key exemplars is the classification of 'race', especially in Apartheid South Africa. The racial classification created, as legal entities, those racialised types, which in turn drove and informed the State's racialised politics and the mechanisms of social exclusion. In a parallel way, successive manuals of medical diagnosis can create the medical conditions they describe—notably but not exclusively in the field of mental health—and in doing so, they actively construct the medical conditions they describe. Such diagnostic categories reflexively determine processes and outcomes of diagnosis and treatment of those who are so classified.

The critique of classification has been most visible in the sociological treatment of data from official sources. Official statistics and rates are gathered on the basis of categories that are applied on the basis of decision-making processes. Among the classic topics for such critical analysis are: suicide (J. M. Atkinson, 1978); types of crime (Sudnow, 1965); educational ability (Cicourel & Kitsuse, 1963). They and similar studies were incorporated into the critique of 'official statistics' that was associated in part with the movement for radical statistics.

The reflexivity of method and measurement impinges directly on the worlds of university research and of education. Universities are subject to multiple versions of 'measurement' that create definitions of success, and that arc back on the behaviour of academic researchers themselves. Periodic national assessments of research performance, such as the Research Excellence Framework (REF) in the UK, and now conducted in one way or another in other national systems too. The published criteria define categories of excellence for published work, with score that are combined with other measures, such as research grant income, numbers of graduate students, and 'research environment'. These generate numerical grades that are widely used to generate league tables of excellence. In the first place, the national criteria, or standards (cf. Bowker & Star, 2000) construct their own measure of excellence. The reflexivity of such a grading system means that it in turn affects the behaviour of academics, individually and collectively, who organise and publish their research in order to optimise their 'scores'.

A similar example from the world of education is given by the PISA global comparison of national levels of educational attainment. That exercise is based on readily quantifiable—and hence comparable—data. Consequently, attainment is defined in terms of a restricted range of quantified measures, based on standardised tests. The average of scores across all school students who participate is then taken as an indicator of national success in educational performance. Now the reflexive influence on behaviour derives from the fact that in the attempt to improve their comparative performance, teachers and policy-makers devote more time, attention and resources to the sort of skills that are enshrined in the assessment regime. This phenomenon is widely known as 'teaching to the test'. Hence, the methods of inquiry create the kinds of competence that are made to stand proxy for the success or failure of entire education systems. As with the periodic assessment of university research, PISA scores are used to construct league tables. And such rank ordering reflexively constructs its own version of reality. Sober inspection of many league tables reveals that, even on the basis of the restricted range of chosen measures, many nations or universities are in fact very similar to one another. But ordering them into a single ordinal measure emphasises *difference*, and even small differences in outcomes can be amplified into significant distinctions. That in turn can redouble efforts to improve one's standing by modifying behaviour in accordance with the method of measurement. This form of reflexivity can be identified across an increasing number of arenas in public life, where performance is measured and league tables are constructed.

Our consideration of methodological reflexivity is here illustrated with just some pertinent examples. It could be extended more or less indefinitely. The topic reminds us that attention to methods and measurements should be a recurrent interest for the social sciences. Such a collective concern does not translate into formulaic prescriptions equivalent to introductory 'how to do it' textbooks. Rather, it implies a commitment to critical methodological analysis that is not satisfied by strident celebrations of 'postpositivism', 'postmodernism' and the like.

Positional Reflexivity and Reflexivity of Membership

In relatively recent years, positionality has become something of a byword among social scientists—especially those who conduct qualitative research of various sorts. It marks the recognition that the researcher's personal characteristics can have consequences for social research at various levels: unexamined assumptions about social life; blind-spots concerning one's self and others; opportunities for empathetic understanding. It also recognises that knowledge-production in the social sciences necessarily implicates the researcher in a nexus of power—not merely power over the researched, but also the power to define the proper nature of research itself: problems and topics, research strategies, preferred outcomes and modes of representation. There is equally the power to exclude topics from scrutiny.

A concern for positionality and reflexivity can be seen as a way to transgress dominant forms of knowledge and inquiry, related to other bases for such critical perspectives. For instance:

> Feminism and the so-called postmodern turn in the social sciences represent a serious challenge to the methodological hegemony of neo-positivist empiricism. One of the main attractions of 'traditional' neo-positivist methods is that they provide a firmly anchored epistemological security from which to venture out and conduct research.... Years of positivist-inspired training have taught us that impersonal, neutral detachment is an important criterion for good research. In these discussions of detachment, distance, and impartiality, the personal is reduced to a mere nuisance or a possible threat to objectivity. The threat is easily dealt with. The neopositivist professional armor includes a carefully constructed public self as a mysterious, impartial outsider; an observer freed of personality and bias. (England, 1994: 82)

It is clearly not just a matter of the researcher's social and personal position(s).The reflexive process of research engagement means equally that the position(s) of the research 'others' is of equal significance. England, for instance, researched marginalised groups and expresses her consciousness of the dialectical relationship between the 'positions' of researcher Self and researched Other. Reflexivity of positionality, therefore, is not just about individual researchers, or even their research community. The positioning of 'others', such as the hosts of fieldwork, is equally important. In a discussion of 'home' and 'field' in anthropology, for instance, Knowles (2000) suggests that 'In order to understand the relationship between home and field, we thus need to examine the researcher's intellectual, political and transnational autobiography' (p. 56l) in order to comprehend the ethnographer's sense of home as a resource for making sense of Others. It is the dialectical relationship between self and others, between home and field, that informs the ethnographic imagination.

Rose (1997) offers an influential overview of feminist standpoint and reflexivity, exploring their complexity and paradoxes. Writing from within feminist geography, she notes that many scholars, such as Haraway (1991), have advocated feminist standpoint approaches, and have tied it to the issues of reflexivity. She succinctly examines some of the difficulties and paradoxes inherent in a programme of what she refers to as *transparent reflexivity*. By that she means the desire to map exhaustively and take account of, multiple forms of power. She suggests that:

> The visible landscape of power, external to the researcher, transparently visible and spatially organized through scale and distribution, is a product of a particular kind of reflexivity, what I will call 'transparent reflexivity'. It depends on certain notions of agency (as conscious) and power (as context), and assumes that both are knowable. (p. 311)

But claiming to know and to take account of the entire range of influences, seems paradoxically to mirror the kind of authorial omniscience that 'reflexivity' challenges. The implicit claim to omniscience Haraway refers to as 'the god-trick'. As Rose herself continues, researchers may be called upon to account for themselves in terms of how their work confronts power, to account for and justify their fields of study, but

> … these analytical claims are little different from the god-trick Haraway—and many feminist geographers—have critiqued so thoroughly. Feminist geographers have certainly situated their analytical gaze, and are now staring hard from locations in the material histories of inequality. But this positioning is still producing some very thorough demands for knowledge. … Indeed, the answers are so massive, the questions are so presumptuous about the reflective, analytical power of the researcher, that I want to say that they should be simply unanswerable: we should not imagine we can answer them. For if we do, we may be performing nothing more than a goddess-trick uncomfortably similar to the god-trick. (p. 311)

It is clear that 'positions' are not fixed, determinate entities, and that one cannot enact reflexivity through a check-list or matrix of given characteristics. It is clearly relatively easy to think in terms of a fairly rigid topography of 'positions'—gender, sexual orientation, ethnicity, age, social class. One can readily create what are essentially stereotypes: the elderly male straight white professor, or the feminist lesbian early career researcher. But such a matrix of positions is not in itself sociologically or anthropologically sensitive or nuanced. Each is enacted, each is performed, each is itself located within a matrix of social relationships. Positional reflexivity does *not*, therefore, reduce the issue to matters of individualised biography. It does, however, recognise that the researcher's identity has implications for all the kinds or levels of reflexivity we have identified hitherto. Positionality here implies that issues of gender, ethnicity and status are all implicated in the research process. But such positionality should not be understood in mechanistic terms. One's ethnicity or sexual orientation (say) may reflexively interact with one's disciplinary commitments in leading us towards framing researchable phenomena and following particular methods. Positionality may lead us towards feminist standpoint perspectives, or queer theory, or critical race theory. Equally, however, such reflexive framing of research should not be regarded as deterministic, any more than the other dimensions of reflexivity we have identified hitherto. It is, after all, axiomatic for anthropologists and sociologists—ethnographers in particular—to regard identities as mutable (at least to some degree). It would, therefore, be a failure of intellectual nerve to assume that one's gender, ethnicity, nationality or sexual orientation would exert a determining effect. It is the work of the social scientist to understand and navigate cultural boundaries, and therefore she or he should be capable of achieving some degree of freedom from biographical constraint. This is not an apology for hegemonic perspectives in science and social science. But equally, a critique of white male epistemic privilege is not secured simply by invoking a contrasting social category.

Clearly positional reflexivity relates directly to standpoint theorising, in turn most closely associated with feminist scholarship. Delamont (2003) provides and extensive discussion of this, from the perspective of feminist sociology. Stanley and Wise (1993) provide a classic exposition. But we must recognise that Stanley and Wise have a more subtle argument than a simple appeal to 'position' might seem to imply. In fact, they suggest that the feminist standpoint provides a way of making everyday social life 'strange'. It is, therefore an idea informed by phenomenology and the ethnomethodological perspective on making everyday phenomena 'anthropologically strange'. Ideally, therefore, standpoints in all their varieties can furnish multiple interpretative frames for research and inferences. As Nagar (2014: 82) points out:

> A widespread engagement with reflexive practices by feminist ethnographers has generated rich dialogues about the methodological and epistemological dilemmas endemic to fieldwork, as well as the challenges associated with identity politics as they affect academic, interpersonal, institutional, and intellectual relationships. *Such reflexivity, however, has mainly focused on examining the identities of individual researchers rather than on how such identities intersect with institutional, geopolitical, and material aspects of their positionality.* (emphasis added)

Social scientists need to apply their professional expertise, therefore, in avoiding crudely over-simplified versions of social positions. Researchers do not inhabit fixed social roles that over-determine their personal and intellectual orientations. Field research in particular rests on shifting identities, on the part of the researcher and the researched. The successful conduct of ethnography, for instance, implicates the ethnographer and her hosts in a web of variable relationships and mutual evaluations that shift over time. Boundaries and identities are constantly subject to negotiation and adaptation. Social scientists are not permanently fixed as novice doctoral researchers in subordinate positions, but equally members of the professoriate are not guaranteed influence and control either. Even where a researcher is in a relatively junior position, he or she may have very different resources from those available to other actors: the time to carry out the research, to reflect, and to write is a precious resource not shared by the great majority of workers, for instance. Differentials in resources—including cultural capital—cannot be reduced to simple categories of class, race or gender. Nagar goes on to identify an imbalance in current research and publication practice: that qualitative researchers are routinely expected to discuss their positionality and identity, while: '…no such expectation applies to scholars using other research methods (e.g. quantitative, archival, and textual analyses), with the consequence that scholars who do not work within ethnographic or life-history traditions 'do not need to respond to critiques of representation and, at worst [that] results in further marginalization of personal narratives in producing knowledge' (p. 84). This asymmetry reflects deeply entrenched assumptions about differences between 'qualitative' and 'quantitative' research strategies, and the implicit assumption that quantitative work is sufficiently 'objective' that it requires no further scrutiny or self-examination on the part of its practitioners.

To comprehend the unconscious influence of personal and academic biography one needs to exercise tough intellectual work. If it is demanding to make familiar a strange research setting, then it is even harder to make oneself sufficiently 'strange' to render one's positionality the object of research scrutiny. It is, therefore, at this 'level' that the essential nature of research reflexivity leads towards something more akin to reflection or even introspection. Biographical reflexivity is a particular issue, given the distinctive nature of social research, and of interviewing and ethnography in particular. The observation entailed in participant observation passes in two directions, in the mutual interrogation of the ethnographer and her or his hosts. This goes beyond the reflexivity of most natural sciences, where the phenomena are framed by the observer (scientist) but that observer is not subject to the reverse gaze. The fieldwork of ethnography is predicated on interactions, encounters and conversations. It is by definition dialogic, at least in that limited sense. Consequently, biographical reflexivity is partly determined not just by the ethnographer's ascribed characteristics, but also by the host's perceptions and judgements of her/him. The ethnographer's self-presentation and the reciprocal moral evaluations influence the nature and scope of data collection, as well as pervading the ethnographer's interpretative frameworks.

In one sense, of course, all the types of reflexivity we identify relate to the researcher's *membership*: her or his membership of an academic discipline or a national tradition, and membership of a community of research practices. Here, by contrast, we again refer to a more restricted and specific kind of membership. That is, membership of a distinctive community of practice, or an occupational group, or a leisure activity. Such research and its distinctive characteristics is often encountered among members of professions who undertake academic research on their own occupation's members: teachers conducting educational research; nurses on nursing. That in turn bears directly on the fundamental basis of ethnographic knowledge. Generically, the issue of membership reflexivity relates to the pervasive issue of strangeness and familiarity. The two are constantly in tension in the conduct of ethnographic research. Ethnographers are constantly enjoined to 'make the familiar strange and the strange familiar', and the relationship between strangeness and familiarity is a recurrent one in the methodological literature on ethnographic research (Delamont & Atkinson, 1995, 2021).

Now in practical terms, the reflexivity of membership derives from the possibility of ethnographers studying aspects of their 'own' culture. Admittedly, this is a slippery notion, and it is often invoked without due care. We cannot assume that any 'Western' sociologist or anthropologist occupies 'the same' cultural space as any other member of Western society. There is no measure of cultural distance that would allow us to engage in the necessary calculus. It is clearly absurd to suggest that a white, English well-educated male has much cultural capital in common with crack-cocaine users in South London (Briggs, 2013). The fact that Rhys-Taylor lives in London, where he studied the sensory worlds of cuisines, such as fried chicken or bush-meat, does not mean that he and the local people automatically share a common culture, or even equivalent social spaces (Rhys-Taylor, 2018). In both of these studies, and many like them, the ethnographer's interpretative competence is not 'given', but is achieved as a consequence or practical, intellectual and imaginative *work*.

So the notion of 'membership' and cultural difference/similarity must be used with some care. Nonetheless, there is a tradition of ethnographic work among the ethnographer's 'own' or 'home' culture. By that we mean something much more specific than just an English social scientist working in an English setting, say, or a Scot studying a Scottish community. Rather, we are referring to that style of fieldwork whereby the researcher deliberately attempts to comprehend a culture in which she or he is already well-versed, a social group or community of which s/he is already a member. Indeed, this is one of the original usages of the term *autoethnography* (Hayano, 1982). When Hayano studied poker-players, being a poker-player himself, he was explicitly studying a social milieu of which he was already a member, and with which he was familiar. He called it 'autoethnography'. Usage of that term has subsequently changed quite radically to mean something more like treating oneself as the central focus of the research, and in that later sense autoethnography has become a genre in its own right, which owes more to autobiography or autofiction than the detailed ethnographic exploration of a sub-culture with which the author is familiar (Atkinson, 2020). Autoethnography, in that latter sense, of writing primarily from the researcher's own point of view, has become widespread among qualitative researchers.

Despite the popularity of recent autoethnographic texts, the practice is not all that novel. There have in fact been many studies in which the ethnographer has studied her or his 'own' setting. There were plenty of studies based on the author's own biography: Becker (1951) on jazz musicians, Scott (1968) on cultures of horse-racing, Davis (1959) on cab-drivers are among the American classics. More recently, there have been many studies by researchers who were or had been practitioners in some specialised domain—such as education, nursing or social work. Such research 'at home', or at least near-to-home suggests a particular form of reflexivity. The topic has been argued at length, as Delamont and Atkinson (2021) put it, it is a matter of *fighting familiarity*. That is, the intellectual imperative to bracket or suspend one's own taken-for-granted cultural categories and treating as 'anthropologically strange' one's chosen research problems and settings. Fighting familiarity also means addressing one's familiar research strategies. In the absence of reflective critique of a research strategy and its apparatus researchers of all persuasions run the risk of ritualistically replicating what they already 'know' to be the case.

In the current methodological climate autoethnography has not necessarily led to an enhanced understanding of 'the other', nor sustained attempts to disrupt the familiarity of a research setting. Autoethnography too readily becomes a form of 'narcissistic reflexivity', a mode of introspection that becomes solipsistic. In its contemporary guise, autoethnography is a genre of autobiographical exploration. It is a manifestation of a contemporary culture of confession and self-revelation. As a consequence, it is antithetical to the general implications of epistemic reflexivity, in the absence of collective knowledge-production and the presence of an uncritical, appreciative celebration of the author's selfhood.

Knowledge, expertise and experience can directly impinge on the formulation of research perspectives and the interpretation of ethnographic exploration. The competent practitioner who wishes to study their own place of work or their own community of practice can frequently find local activities too familiar to be able to 'see' potentially researchable phenomena. The observational filter is that of her or his taken-for-granted competence. The ethnographer studying their 'own', therefore, can find it hard to look through or beyond (no such metaphor is quite right) the professional and habitual frames of reference. That is not an insurmountable problem, and it is not always negative. We shall return to the consequences of socialised competence below, in a brief discussion of *unique adequacy*.

Hence, reflexivity of membership tends to frame ethnographic knowledge that is couched in terms of pre-given concepts and research problems. It is easy to engage in professional judgements rather than anthropological or sociological ones. The educator who studies education, the nurse who studies nursing care, or the police officer who studies policing thus needs to work hard to suspend taken-for-granted assumptions. What is required is often a kind of estrangement—a form of *phenomenological reduction*—whereby the ethnographer needs to suspend or bracket out her or his accustomed, sometimes professional, assumptions and expert knowledge. This is not because the ethnographer is some sort of feral ignoramus, who knows nothing and learns little, but in order to confront the Others and their culture afresh: as if they were comparative strangers.

Estrangement confers imaginative and analytic distance, and epistemic distance leads towards the freshness of the ethnographic gaze. Equally, of course, familiarity can confer some advantages for the would-be ethnographer. A working knowledge of esoteric techniques and skills can convey a distinctive perspective on, say, scientific laboratories, or medical practice. Some commentators have advocated what they call unique adequacy as a prerequisite for the close study of such practices. Here we neither advocate its necessity nor criticise such an approach (but see Atkinson & Morriss, 2017). Rather we draw attention to the fact that members' familiarity or strangeness reflexively impinge on how research questions and problems are framed, what intellectual and imaginative work is brought to bear, and possible limitations on what is observable or thinkable ethnographically.

Reflexivity of membership illustrates the more general point quite forcibly. It is not a matter that can be resolved, there is a constant ethnographic tension between familiarity and strangeness. It is also manifested in current claims concerning *Indigenous research*: for major expositions of that movement, see Denzin et al., 2008; Smith, 2021; Kovach, 2010a). Claims for Indigenous research go well beyond issues of reflexivity and knowledge alone, and they also extend beyond ethnographic research. It is, however, argued by its advocates that membership of a particular 'indigenous' group furnishes privileged understanding. There are a number of challenges surrounding this set of claims (see Ryen & Atkinson, 2016). But there is no doubt that the generic reflexivity of membership extends to ethnic or cultural membership. The current literature is mostly confined to a very limited number of ethnic groups (in Australia, New Zealand and the Pacific, and in North America especially), while there are many possible 'indigenous' people. On the other hand, the entire historical thrust of anthropology and sociology must surely render categories like 'Indigenous' highly problematic. In the same way, assumptions about culture and identity that are based on essentialising categories are antithetical to the ethnographic inspiration. Consequently, the reflexivity of membership raises significant issues concerning cultural knowledge, field relations or roles, and the taken-for-granted assumptions that are brought to bear in ethnographic fieldwork and analysis.

Advocates of Indigenous Research can sometimes be guilty of perpetuating stereotypes concerning research practice that do not lead to a subtle analysis of reflexivity. For instance, Porsanger (2004: 106) deploys a common trope in promoting it:

> In the Western understanding research in general may be defined as an investigation or experiment aimed at the discovery and interpretation of facts. Research includes collecting information about a particular subject, revising accepted theories or laws in the light of new facts, and the practical application of such new or revised theories or laws.

An inappropriate emphasis on 'science' is part of this stereotype: 'Processes of theorizing and measuring what is considered to be "scientifically acceptable" have been based on Western philosophy and imply a notion of objective research' (Porsanger, 2004: 110). Likewise, Wilson (2008: 176–177) asserts that,

> One major difference between the dominant paradigms and an Indigenous paradigm is that the dominant paradigms build on the fundamental belief that knowledge is an individual entity: the researcher is an individual in search of knowledge, knowledge is something that is gained, and therefore knowledge may be owned by an individual.

This is then developed as:

> An Indigenous paradigm comes from the fundamental belief that knowledge is relational. Knowledge is shared with all of creation. It is not just interpersonal relationships, not just with research subjects I may be working with, but it is a relationship with all of creation. (ibid.)

Members of research teams, especially in the natural sciences, would be startled by the assertion that research-based knowledge is always regarded as individually owned, but that is a minor detail compared with those sweeping generalisations. While the interests of Indigenous researchers are of paramount importance, they are not well served by stereotyped views, either of 'Western' scholarship or of Indigenous cultures. The principle of epistemic reflexivity acknowledges that knowledge-production is grounded in multiple relationships and engagements that encompass the researcher and the research participants, the researcher's community of practice, and networks of technique and method. It is not something that is special or unique in Indigenous Research. Equally, an acknowledgement of epistemic reflexivity ought to recognise that all research is 'relational'. It is the collective task of all epistemic communities to explore and evaluate *how* the multiple relations (disciplines, networks, biographies) work together, or work against each other—in the collective work of knowledge-production.

The most important issue for us is this: does the reflexivity of membership in and of itself guarantee authentic and privileged understanding of one's 'own' culture? Moreover, does the thrust of Indigenous Research adequately address the complexities of positional and membership reflexivity? And does the call for Indigenous Research unduly constrain the intellectual work of cross-cultural exploration? The overall claim is that there are modes of knowledge and experience that are distinctive to Indigenous peoples. This may include: orientations to the natural world, matters of spirituality, the appreciation of collective activity, and an emphasis on shared oral traditions expressed through talk and narratives (Kovach, 2010b; Wilson, 2001, 2008). That is true of many cultural settings not all of which are marginalised and Indigenous. The claim that only Indigenous researchers who share a common heritage with the people among whom they work can undertake such research is more contentious. This is not the place to address all of the possibilities and limitations of Indigenous Research. We raise it here because it is a special case of positionality and membership, and it throws them into relief. The Indigenous Research movement proposes that research can and should be conducted only by members of the relevant ethnic community, only with the express collective permission of that community, and undertaken on behalf of that community. A pressing and unresolved methodological issue here is the extent to which a 'member', who claims that their membership grants a particular degree of legitimacy, can actually treat the host culture as anthropologically strange.

Membership and positionality raise some significant issues for the conduct of ethnography. If the ethnographer is—as s/he implicitly claims—able to understand the social world from the point of view of the Other(s), can s/he not also shift 'position'? If the ethnographer can be enough of a shape-shifter to cope in 'the field', can s/he not also be an intellectual shape-shifter too? In other words, it would be ethnographically inadequate to regard social categories and identities as fixed.

Identities are seen as fluid, plastic, mobile. Is the ethnographer therefore always in a fixed 'position'? It is an essentialist mistake to imply that no male researcher can comprehend a woman's perspective. His taken-for-granted categories of thought can be interrupted and disrupted; his viewpoint could shift radically; new phenomena could open themselves for investigation. The same is true for female researchers, self-evidently. Indeed, good research ought to bring such disruptions in its train. Of course, that hypothetical ethnographer might be criticised for appropriating ideas that are 'owned' by other scholars. But the intellectual exercise—without exaggerated claims for identity—could potentially create new spaces and resources for inquiry.

This is not quite the same approach as attempting to tell 'the same' story from different actors' points of view—the Rashomon effect—as essayed by Wolf (1992) in her 'thrice-told tale'. The attempt to shift narrative point of view is, in itself, a challenging exercise in intellectual distancing, but it is not the practice we have been alluding to here. In practice, the exercise would likely, and perhaps justifiably, fall foul of identity politics. And if carried out textually, it might result only in pastiche of different intellectual styles. Nonetheless, the thought-experiment could be an illuminating one: an exercise in further estrangement and observer-alienation. We turn to such textual considerations in the next section.

Textual Reflexivity

As we pointed out earlier in our comments on Barad and Woolgar, reflexivity is not only a matter of textual forms and representations. But representation is a significant aspect of reflexivity in its own right. It is especially central to the creation and reception of ethnographic texts, in anthropology, sociology and related fields, where textual representation has been a recurring concern for many years (Atkinson, 1990, 2020). The issue relates to a much broader interest in texts and representations of scholarly and scientific work. We have already made reference to the reflexive turn in studies of science and technology (STS), which seems to have been a relatively short-lived—if significant—period in that discipline (e.g. Ashmore, 1989; Woolgar, 1988). Because STS has tended to be somewhat inward-looking, 'reflexivity' in that context was sometimes taken to be limited to matters of textual forms and conventions: see for instance Haraway (1997). It was, however, an aspect of a broader intellectual movement, sometimes glossed as the 'rhetoric of inquiry' programme (Nelson et al., 1987). That constellation of interests covered the textual forms of the natural, social and cultural disciplines. Its authors displayed the *conventions* of academic writing and representation characteristic of selected disciplines.

Analyses of representations and texts were by no means confined to the most 'literary' of disciplines. McCloskey (1985), for example, examined how as a discipline Economics—for all its 'scientific' appearance—deployed distinctive rhetorical resources in constructing its arguments. Typically, McCloskey suggested, economists deploy a metaphor (a 'model') or construct a 'story' (a narrative), or both. They construct 'as if' narratives in developing their arguments (see also McCloskey, 2019). Even the scientific paper obeys textual conventions in establishing its implicit claims to impersonal factual status. The scientific journal paper is now so thoroughly evolved as a genre that it may seem the 'natural' way to report scientific research and findings. But both framing the paper and embedded within it is a variety of persuasive rhetorical devices (e.g. Yearley, 1981). Likewise, Bazerman (1981) explored the rhetorical devices and assumptions of texts in three different disciplinary genres, showing how they varied in constructing their arguments and mobilising their persuasive devices.

The reflexivity of written accounts is an especially pertinent topic for ethnographic and other qualitative inquiry. The construction of an ethnographic monograph is an egregious example of the textual, compositional work implied by ethnographic work. It reminds us that 'the ethnography' refers simultaneously to the research project and to the resulting text. The two imply one another. So while *all* science and scholarship draw on rhetorical, textual conventions, ethnography is especially dependent on its own textual devices. The 'field' of fieldwork is created partly by the practical methods of fieldwork, and partly by how it is written into recognisable and credible reality (Atkinson, 1990).

In the 1980s anthropology, especially American cultural anthropology, underwent a collective conversion to an awareness of textual reflexivity. The most prominent and influential publication of that period was the collection *Writing Culture*, edited by Clifford and Marcus (1986). It was by no means the first publication to examine anthropological ethnography as text. But it appeared at a particular moment in the discipline's trajectory, when collective concern was focused particularly on the authority of anthropology as a discipline, and on the knowledge-claims of individual anthropologists. A once supremely confident discipline was questioning its collective identity and its foundations. The collection of essays was identified in many quarters as helping to precipitate a 'crisis of representation'. It was argued by the editors and their authors that the textual style of the traditional anthropological monograph inscribed a characteristic stance on the part of the anthropologist. Anthropological monographs characteristically implied an image of the ethnographer-author as an omniscient observer, who was simultaneously invisible. Their presence in the field and as author of the monograph was marked only by absence. The text was constructed, therefore, by an invisible author who rarely—if ever—appeared as an actor in the field. The impersonal, omniscient author, it was argued, was matched by a text that self-confidently reconstructed 'the culture'. It admitted of little or no doubt or hesitation, and rarely did justice to the complexities, inconsistencies and contradictions of culture. The critique also reflected collective unease about the colonialist legacy of anthropology's disciplinary origins. The single point of view and the semantically unmarked, authoritative author seemed to project a Western, predominantly male, hegemonic authority. In contrast, calls were voiced for ethnographic texts that were more 'open', even 'messy', and polyvocal. In other words, the textual reflexivity of ethnographic writing would reflexively construct a very different version of 'the field'. The field itself would be reconstituted in ways that portrayed a multiplicity of perspectives and voices, not a single homogeneous picture, unified through a single authorial voice. This general perspective on textual reflexivity (though not by that name) was reflected in further collections of essays that extended and developed the basic arguments.

To the original critique enshrined in *Writing Culture* was added a distinctively feminist perspective (Behar, 1993; Behar & Gordon, 1995). The omniscient ethnographer implied in the text was gendered. Although, of course, there had been major female anthropological authors from the earliest days of the discipline (Mead, Benedict, Powdermaker, Mair, Richards), the analytic gaze of the anthropological monograph could be identified as predominantly one of a male, white hegemonic mentality. Consequently, the use of conventional textual formats in reporting 'other' cultures became thoroughly implicated in broader debates about the proper conduct of anthropology and analytic perspectives that anthropologists bring to bear on their subject-matter.

On the other hand, the actual devices and textual machinery of reflexivity were not fully addressed in those polemical texts. That was better represented in the work of Geertz (1988), whose close reading of a small number of classic anthropologists illuminated how their written style reflexively constructed each one's distinctive reconstruction of their chosen research field. It made perfect sense for Geertz to examine the written styles of his discipline, given that his overall approach was to treat cultural analysis in terms of textual interpretation. He deals specifically with Malinowski, Evans-Pritchard, Benedict and Lévi-Strauss. Each of those anthropologists displayed a distinctive written style, reflecting an amalgam of national and disciplinary context, personal style, and analytic preoccupations. In doing so, the author reflexively constructed her or his chosen social milieu. Each monograph displays its origins, in terms of time, place and intellectual field. Each reflects an authorial style. As a consequence, monographs differ within the canons of disciplinary conventions. There is, therefore, a dialectical relationship between the author's 'signature' and the exposition of ethnographic fact. About the general issue of ethnographic authorship, Geertz writes on the necessity to scrutinise how anthropologists author-ize their monographs:

> The difficulty is that the oddity of constructing texts ostensibly scientific out of experiences broadly biographical, which is after all what ethnographers do, is thoroughly obscured. The signature issue, as the ethnographer confronts it, or as it confronts the ethnographer, demands both the

> Olympianism of the unauthorial physicist and the sovereign consciousness of the hyperauthorial novelist, while not in fact permitting either. The first brings charges of insensitivity, of treating people as objects, of hearing the words but not the music, and, of course, of ethnocentrism. The second brings charges of impressionism, of treating people as puppets, of hearing music that doesn't exist, and, of course, of ethnocentrism. Small wonder that most ethnographers tend to oscillate uncertainly between the two, sometimes in different books, more often in the same one. Finding somewhere to stand in a text that is supposed to be at one and the same time an intimate view and a cool assessment is almost as much of a challenge as gaining the view and making the assessment in the first place. (p. 10)

It also small wonder the anthropologist's authorial style (signature) is of considerable moment, and has come to occupy a central place in anthropological debate.

From within sociological ethnography, textual reflexivity has received particular attention, from authors approaching the topic in a manner parallel to but independently of the anthropological tradition. Van Maanen (1988, 2011), for instance, documented distinctive styles or genres of ethnographic text. Such genres construct fieldwork, the field and the fieldworker in contrasting ways. The realist genre is the style of most conventional ethnographic reporting. As Atkinson (1990) demonstrated, 'realist' ethnographies characteristically depend on various textual methods in order to construct the verisimilitude of their accounts, as well as characteristic tropes, such as metaphor and irony, in order to create sociological significance. Van Maanen also identified a confessional style, that ethnographers have used to explore the personal aspects and commitments of their own fieldwork. Confessionals have often explored the vicissitudes of ethnographic fieldwork, the personal and even emotional effects on the ethnographer. To that extent, such accounts acknowledge the reflexive nature of ethnographic research, although in our terms, they approximate more to the category of 'reflective' texts—closely related to but not identical to reflexivity in research. At the time of Van Maanen's first publication, ethnographic monographs and confessional accounts were usually separate texts. Indeed, it was suggested that the separation allowed the ethnographer to express the 'personal' side of fieldwork without any possible corruption of the realist text, the latter being regarded as the authentic version of the research. In more recent years, however, those genres have become intermingled, as aspects of confessional and autobiographical writing have permeated many ethnographic texts. The conventions of realist reporting have not merely been identified and discussed: they have been challenged and, in some quarters, replaced with other conventions. It is important to recognise that we are always in the realms of textual convention, as there is no one style of ethnographic writing that is a more fundamentally authentic or faithful mode of representation or reconstruction.

The field of qualitative and ethnographic writing is now populated by a multiplicity of styles, that singly and together challenge the conventions of realist writing and reading. Those genres include ethnographic fiction, polyvocal texts or ethno-dramas, and academic 'poems'. What they share is the author's desire to disturb the taken-for-granted assumptions of the ethnographic monograph or journal paper. They draw attention to the textual work of reconstruction and representation. Whereas the 'classic' ethnographic monograph usually presented an assured authorial voice, writing from a perspective based on intellectual authority, many contemporary texts deliberately seek a more tentative voice, in which the relative certainty of the ethnographic text is replaced by a more uncertain, fragmented narrative.

References

Amit, V. (Ed.). (2000). *Constructing the field: Ethnographic fieldwork in the contemporary world*. Routledge.

Ardener, E. (1972). Belief and the problem of women. In J. Lafontaine (Ed.), *The interpretation of ritual* (pp. 135–158). Tavistock.

Ardener, E. (1975). The 'problem' revisited. In S. Ardener (Ed.), *Perceiving women* (pp. 19–27). Dent.

Ashmore, M. (1989). *The reflexive thesis: Wrighting sociology of scientific knowledge*. University of Chicago Press.

Atkinson, J. M. (1978). *Discovering suicide: Studies in the social organization of sudden death*. Macmillan.

Atkinson, P. (1990). *The ethnographic imagination: Textual constructions of reality*. Routledge.
Atkinson, P. (2009). Ethics and ethnography. *21st Century Society: Journal of the Academy of Social Sciences, 4* (1), 17–30.
Atkinson, P. (2020). *Writing ethnographically*. Sage.
Atkinson, P., & Housley, W. (2003). *Interactionism*. Sage.
Atkinson, P., & Morriss, L. (2017). On ethnographic knowledge. *Qualitative Inquiry, 23*(5), 323–331.
Atkinson, P., & Silverman, D. (1997). Kundera's *Immortality*: The interview society and the invention of the self. *Qualitative Inquiry, 3*(3), 304–325.
Babcock, B. (1980). Reflexivity: Definitions and discriminations. *Semiotica, 30*, 1–14.
Bauman, R. (1986). *Story, performance and event*. Cambridge University Press.
Bauman, R. (1989). Performance. In E. Barnouw (Ed.), *International encyclopedia of performance* (pp. 262–266). Oxford University Press.
Bazerman, C. (1981). What written knowledge does: Three examples of academic discourse. *Philosophy of the Social Sciences, 11*(3), 361–387.
Becker, H. S. (1951). The professional dance musician and his audience. *American Journal of Sociology, 57*(2), 136–144.
Behar, R. (1993). Women writing culture: Another telling of the story of American anthropology. *Critique of Anthropology, 13*(4), 307–325.
Behar, R., & Gordon, D. (Eds.). (1995). *Women writing culture*. University of California Press.
Berger, H. M., & Del Negro, G. P. (2002). Bauman's verbal art and the social organization of attention: The role of reflexivity in the aesthetics of performance. *The Journal of American Folklore, 115*(155), 62–91.
Bowker, G., & Star, S. L. (1999). *Sorting things out: Classification and its consequences*. MIT Press.
Bowker, G., & Star, S. L. (2000). Invisible mediators of action: Classification and the ubiquity of standards. *Mind, Culture and Activity, 7*(1–2), 147–163.
Braun, V., & Clarke, V. (2019). Reflecting on reflexive thematic analysis. *Qualitative Research in Sport, Execise and Health, 11*(4), 589–597.
Briggs, D. (2013). *Crack cocaine users: High society and low life in South London*. Routledge.
Bruce, D. B., Jr. (1992). W E B Du Bois and the idea of double consciousness. *American Literature, 64*(2), 299–309.
Charmaz, K. (2014). *Constructing grounded theory* (2nd ed.). Sage.
Cicourel, A. V., & Kitsuse, J. I. (1963). *The educational decision-makers*. Bobbs-Merrill.

Clifford, J., & Marcus, G. E. (Eds.). (1986). *Writing culture*. University of California Press.

Criado, T. S., & Estallella, A. (2018). Introduction: Experimental collaborations. In A. Estalella & T. S. Criado (Eds.), *Experimental collaborations: Ethnography through fieldwork devices* (pp. 1–30). Berghahn.

Davies, C. A. (2008). *Reflexive ethnography: A guide to researching selves and others* (2nd ed.). Routledge.

Davis, F. (1959). The cabdriver and his fare: Facets of a fleeting relationship. *American Journal of Sociology, 65*(2), 158–165.

Delamont, S. (1989). *Knowledgeable women*. Routledge.

Delamont, S. (2003). *Feminist sociology*. Sage.

Delamont, S., & Atkinson, P. (1995). *Fighting familiarity*. Hampton Press.

Delamont, S., & Atkinson, P. (2021). *Ethnographic encounters*. Routledge.

Denzin, N.K. (2013). Coda: The death of data? *Cultural Studies↔Critical Methodologies, 13*(4), 353–356.

Denzin, N. K., Lincoln, Y. S., & Smith, L. T. (Eds.). (2008). *Handbook of critical and indigenous methodologies*. Sage.

England, K.V.L.(1994). Getting personal: Reflexivity, positionality, and feminist research. *The Professional Geographer, 46*(1), 80–89.

Estalella, A., & Criado, T. S. (Eds.). (2018). *Experimental collaborations: Ethnography through fieldwork devices*. Berghahn.

Evens, T., Handelman, D., & Roberts, C. (2016). Introduction. In T. Evens, D. Handelman, & C. Roberts (Eds.), *(2016) Reflecting on reflexivity: The human condition as an ontological surprise* (pp. 1–20). Berghahn.

Garfinkel, H. (1967). *Studies in ethnomethodology*. Prentice-Hall.

Geertz, C. (1988). *Works and lives: The anthropologist as author*. Polity.

Gubrium, J. F., & Holstein, J. A. (2003). From the individual interview to the interview society. In J. F. Gubrium & J. A. Holstein (Eds.), *Postmodern interviewing* (pp. 21–49). Sage.

Hammersley, M. (2010). Reproducing or constructing? Some questions about transcription in social research. *Qualitative Research, 10*(5), 553–569.

Haraway, D. (1991). *Simians, cyborgs and women: The representation of nature*. Free Association Books.

Haraway, D. (1997). *Feminism and technoscience*. Routledge.

Hayano, D. M. (1982). *Poker faces: The life and work of professional card players*. University of California Press.

Housley, M., & Smith, R. J. (2011). Telling the CAQDAS code: Membership categorisation and the accomplishment of 'coding rules' in research team talk. *Discourse Studies, 13*(4), 417–434.

Knowles, C. (2000). Here and there: Doing transnational fieldwork. In V. Amit (Ed.), *Constructing the field: Ethnographic fieldwork in the contemporary world* (pp. 54–70). Routledge.
Kovach, M. (2010). *Indigenous methodologies: Characteristics, conversations, and contexts*. University of Toronto Press.
Kovach, M. (2010). Conversational method in indigenous research. *First Peoples Child and Family Review, 5*(1), 40–48.
Mauthner, N. S., & Doucet, A. (2003). Reflexive accounts and accounts of reflexivity in qualitative data analysis. *Sociology, 37*(3), 413–431.
McCloskey, D. (1985). *The rhetoric of economics*. University of Wisconsin Press.
McCloskey, D. (2019). *Economical writing* (3rd ed.). University of Chicago Press.
Mead, G. H. (1934). *Mind, self and society*. University of Chicago Press.
Myerhoff, B., & Ruby, J. (1982). Introduction. In J. Ruby (Ed.), *A crack in the mirror: Reflexive perspectives in anthropology* (pp. 1–35). University of Pennsylvania Press.
Nagar, R. (2014). *Muddying the waters: Co-authoring feminisms across scholarship and activism*. University of Illinois Press.
Nelson, J. S., Megill, A., & McCloskey, D. N. (Eds.). (1987). *The rhetoric of the human sciences*. University of Wisconsin Press.
Pink, S. (2018). Afterword: Refiguring collaboration and experimentation. In A. Estalella & T. S. Criado (Eds.), *Experimental collaborations: Ethnography through fieldwork devices* (pp. 201–212). Berghahn.
Porsanger, J. (2004). An essay about indigenous methodology. *Nordlit*, 15 Special issue on Northern Minorities: 105–120.
Rhys-Taylor, A. (2018). *Food and multiculture: A sensory ethnography of East London*. Bloomsbury.
Rose, G. (1997). Situating knowledges: Positionality, reflexivities and other tactics. *Progress in Human Geography, 21*(3), 305–320.
Ryen, A., & Atkinson, P. (2016). Indigenous research and Romantic nationalism. *Societies, 6*(4), 34. https://doi.org/10.3390/soc6040034.
St. Pierre, E. A. (2011). Post qualitative research: The critique and the coming after. In N. K. Denzin & Y. S. Lincoln (Eds.), *Handbook of qualitative research* (4th ed., pp. 611–626). Sage.
St. Pierre, E. A. . (2015). Practices for the 'new' in the new empiricisms, the new materialisms, and post qualitative inquiry. In N. K. Denzin & M. D. Giardina (Eds.), *Qualitative inquiry and the politics of research* (pp. 75–95). Left Coast Press.
Saldaña, J. (2016). *The coding manual for qualitative researchers* (3rd ed.). Sage.

Schechner, R. (1982). Collective reflexivity: Restoration of behavior. In J. Ruby (Ed.), *A crack in the mirror: Reflexive perspectives in anthropology* (pp. 39–81). University of Pennsylvania Press.
Scott, M. (1968). *The racing game*. Transaction Books.
Smith, L. T. (2021). *Decolonizing methodologies: Research and indigenous peoples* (3rd ed.). Zed Books.
Stanley, L., & Wise, S. (1993). *Breaking out again: Feminist ontology and epistemology*. Routledge.
Strawson, G. (2004). Against narrativity. *Ratio, XVII*(4), 428–452.
Sudnow, D. (1965). Normal crimes: Sociological features of the penal code in a public defender office. *Social Problems, 12*(3), 255–276.
Turner, V. (1982). Dramatic ritual/ritual drama: Performative and reflexive anthropology. In J. Ruby (Ed.), *A Crack in the mirror: Reflexive perspectives in anthropology* (pp. 83–97). University of Pennsylvania Press.
Uprichard, E., Burrows, R., & Byrne, D. (2008). SPSS as an 'inscription device': From causality to decription. *The Sociological Review, 56*(4), 606–622.
Van Maanen, J. (1988). *Tales of the field*. University of Chicago Press.
Van Maanen, J. (2011). *Tales of the field* (2nd ed.). University of Chicago Press.
Whitaker, E. M., & Atkinson, P. (2019). Authenticity and the interview: A positive response to a radical critique. *Qualitative Research, 19*(6), 619–634.
Wilson, S. W. (2001). What is Indigenous research methodology? *Canadian Journal of Native Education, 25*(2), 175–179.
Wilson, S. W. (2008). *Research is ceremony: Indigenous research methods*. Fernwood Press.
Wolf, M. (1992). *A thrice told tale*. Stanford University Press.
Woolgar, S. (1988). *Science: The very idea*. Ellis Horwood.
Yearley, S. (1981). Textual persuasion: The role of social accounting in the construction of scientific arguments. *Philosophy of Social Science, 11*(3), 409–435.

4

Living and Working with Reflexivity

Abstract This concluding chapter returns to some major themes in the contemporary treatment of research reflexivity. It begins by making a distinction between reflexivity and reflection. It is argued that many contemporary usages should be described in terms of reflection or reflective practice, and do not do justice to the full complexity of reflexivity. We end by discussing how we should learn to work with reflexivity and is implications: acknowledging the intellectual craft-work that goes into the production of knowledge.

Keywords Reflexivity · Reflection · Reflective practice · Intellectual craft

Reflexivity and Reflection

It is clear that most, if not all, versions of reflexivity imply some degree of self-consciousness on the part of social researchers. This is, after all, the extension of that sense of reflexivity that social actors are capable of self-awareness of their own actions, as well as those of others. Such human capacity is at the heart of accounts of reflexive modernity, as

E. M. Whitaker and P. Atkinson, *Reflexivity in Social Research*,
https://doi.org/10.1007/978-3-030-84095-2_4

well as generic ideas about the self and its constitution. It is, therefore, necessary to take account of this strand of contemporary usage, while emphasising that it is *not* the sum total of reflexivity and its methodological import. In order to make these distinctions clear, we strongly prefer to write in terms of *reflection* or *critical reflection* rather than reflexivity in this context, bearing in mind that reflexivity is a pervasive feature of social research, while critical reflection is an activity on the part of the researcher. As May and Perry (2017) put it: 'Reflexivity is not just about the ability to think about our actions—that is called reflection—but an examination of the foundations of frameworks of thought themselves' (p. 3). Reflection stems from the researcher's understanding of her or his reflexive engagement with the social world and with the research process that embodies that engagement. As Davies (2008) suggests, '… reflexivity expresses researchers' awareness of their necessary connection to the research situation and hence their effects upon it, what is sometimes called reactivity' (p. 7). In other words, comprehension of research reflexivity leads towards the researcher's capacity for self-awareness and reflection. In accordance with Davies's observation, the essential reflexivity of research stems from the fact that any attempt to describe, classify or otherwise measure phenomena is reactive. There is no point external to the system of researcher, researched, and phenomena from which that system may be surveyed without some degree of reactivity.

From our point of view, *personal* self-awareness invites the cultivation of the researcher as a *reflective practitioner*. The parallel is not exact. When the ideas came to prominence through Donald Schön's monograph (Schön, 1983), they were part of his critique of purely technical-rational thinking among the liberal professions. In contrast, Schön emphasised reflection-in-action, as a practical kind of activity. Reflective practice, pertinently for our discussion here, helps the researcher or practitioner to explore alternative ways of conceptualising issues and problems. Practitioners can, for instance, explore different frames of reference:

> When practitioners are unaware of their frames for roles or problems, they do not experience the need to choose among them. They do not attend

> to the ways in which they *construct* the reality in which they function; for them, it is simply the given reality. (p. 310, emphasis in original)

In much the same way, researcher-practitioners can extend their taken-for-granted models and exemplars. The overall implication is that reflective practitioners can be encouraged to turn their tacitly-held assumptions into objects of professional scrutiny. And they can more readily think in terms of *alternative* interpretative frames, models, precedents, and in general terms make visible their tacit knowledge. This implies going further than just self-congratulatory narcissistic reflection.

Equally wide of the mark are responses that are framed only in terms of the researcher's personal self-consciousness. This is what has sometimes been called 'narcissistic reflexivity'. Myerhoff and Ruby (1982) observed that reflexivity implies reflection, but the two are not the same: '"Reflective" is a related but distinguishable term, referring also to the kind of thinking about ourselves, showing ourselves to ourselves, but without the requirement of explicit awareness of the implications of our display' (p. 3). Many current usages of 'reflexivity' in qualitative research and ethnography are exactly that. They pay insufficient attention to the kinds of epistemic reflexivity that we have outlined, and merely report that the researchers were self-aware or self-critical. The crucial issue is *how* the unavoidable reflexivity of research was worked with and worked through. It is commonplace to find authors claiming to have undertaken 'reflexive ethnography' as a matter of methodological virtue and self-congratulation, where the researcher/author apparently exposes her or his autobiographical origins and predispositions. When this is treated as a personal virtue, insufficient attention is paid to the epistemic and collective aspects of research reflexivity. It is not recognised that the reflexivity of research is what makes it possible in the first place: disciplinary knowledge, methodological means, conventional forms of representation. It provides the essential character of research and is certainly not just the cause of bias and error.

Personal, autobiographical *reflection* undoubtedly has its place in the conduct of research. Reflective practitioners are always to be celebrated. If nothing else, reflection can promote a degree of honesty and transparency in the research process. It becomes dysfunctional, however, when self-examination is transformed into self-regard, or self-absorption. Not least because it collapses all of the intricacies of reflexivity into an individualised act, often amounting to introspection. Madison (2006) makes that point in a polemical fashion: 'I am asking that we reinvigorate our thinking about Otherness relative to ethnography in what feels like a sea of autoethnography mania. I'm feeling the Other... slipping in murky shadows upstaged by a contrived poetics of self' (p. 320). Reflective practice on our engagements with the Others of social research should certainly not be 'upstaged' or excluded by a preoccupation with the Self of the research practitioner.

Stuart (2018) suggests a more nuanced, anthropologically informed, view of the position(s)of the field researcher: 'Once we concede that we are *always* outsiders in at least some fashion, we can begin to alleviate some of the paralyzing thoughts faced by apprentice and veteran ethnographers alike' (p. 216, emphasis in original). Such a perspective invites a recognition that field researchers 'are engaged in an act of transgression each and every time they set foot in the field' (p. 218). In extreme cases, it could result in a narrowing of social research to the virtual exclusion of anything *except* the researcher her/himself. Extreme self-absorption in the name of reflexivity results in the worst excesses of autoethnography, in which the author's own emotional states, interests and experiences substitute for a sustained engagement with a social world. Anderson (2006) makes a clear distinction between 'evocative' autoethnography and its 'analytic' counterpart, arguing that analytic reflection on a social group or setting of which one is a member is a valuable aspect of qualitative social research, while purely autobiographical texts are not. Atkinson (2006), in endorsing Anderson's view makes the case equally strongly, as does Delamont (2009), who criticises what she called the 'narcissistic substitution' of research by self-absorption. The nub of the dispute is not the intrinsic value of self-critique, or conscious reflection on research practice, but the extent to which writing the self has in some quarters got in the way of writing analytically about one's encounters with 'others' in the field.

Living and Working with Reflexivity

There is a danger of allowing a recognition of reflexivity to make research either virtually impossible (intellectual stasis) or nugatory (obsessive self-absorption). But neither is a necessary outcome, both being self-evidently undesirable. In an earlier section we identified an apparent paradox: that reflexivity has been identified as a source of greater objectivity and as a justification for more subjectively personal modes of research activity. We did not go beyond pointing out that this one of several sources of confusion surrounding the ideas of reflexivity itself. While we cannot resolve all of these contradictions or wish away the implications of reflexivity itself, one outcome of our own arguments leads us to suggest that in this context the antithesis of objective and subjective is misplaced. The terminology itself—though deeply embedded in discussions of social research—is unhelpful. The language of objectivity is too often taken to imply that valid knowledge must be based on impersonal, universally applicable technologies of observation, measurement and description. Equally, the language of subjectivity is used to celebrate individualised, unaccountable or experiential inquiry that escapes or transcends the bounds of disciplinary canons of rigour. It should be clear from all we have written that we endorse neither of those extremes.

The very notions of bias and error are problematic in the context of reflexivity. To repeat our basic argument yet again, reflexivity is an inherent feature of research, based on the inescapable fact that our 'data' or our 'findings' are inevitably linked to the nature of our observations and representations. Talk of bias, on the other hand, presupposes a 'true' or 'faithful' representation that is prior to and independent of the methods of investigation and the conceptual apparatus, whereas in the world of research, the phenomena themselves are given form and substance by the means used to study them. Any form of instrumentation or operationalisation helps to define the very objects of that research: attitude scales and measurements of intelligence are obvious examples in the social sciences. Any one method may be open to criticism, of course, and it does not mean that all methods are equally good or fit for purpose. Ours is not yet another manifesto for pure constructivism or methodological relativism. But it does suggest that there exist no pure phenomena that we investigate independently of epistemic, methodological and representational conventions.

How does one 'live with' reflexivity? One lives with it by living through it. In other words, we cannot wish it away, so living with it is precisely what is necessary. What does that entail? Among other things, it demands a thorough recognition of what social research entails. It implies the abandonment of misplaced aspirations based on inappropriate models of 'science'. It also means the recognition that research can never be based on purely neutral methods of inquiry, or media of representation that are purely transparent. It means grasping that the process of research is always the outcome of social transactions and negotiations, that it is thoroughly dependent on ways of writing (and other modes of representation). We have to recognise that such acts of translation are always partial and perspectival. The 'original' from which we translate is itself open to multiple readings, and the language or culture into which we translate does not map perfectly onto the original. Consequently, there can be multiple versions, without accusations of wilful distortion or error. We must recognise that any given translation reflects its time and place, and each carries the traces of its own cultural milieu. This also implies a recognition that ethnography is as much an exercise in cultural and literary work as it is a matter of 'science'. Scholarship requires care, of course, and methodical work (though not the fetishisation of method per se). Reflexivity is never an excuse for arbitrariness or carelessness, justified solely in terms of personal positionality. It *does* mean working with and living with some ambiguity as well as complexity. It also means that the essential reflexivity of research is not something that can be eliminated: the search for a pure, uncontaminated account of the social world is chimerical.

Our view of epistemic reflexivity, in all its guises, also leads towards critical engagements with cherished, often tacit, disciplinary orthodoxies. This is not simply a matter of embracing 'interdisciplinarity', although questioning the boundaries of disciplinary traditions and genealogies should always be undertaken. It also means that just as we try to make the topics of research 'anthropologically strange', we should try to do the same for our intellectual resources, including key concepts and cherished methods. As we pointed out in earlier sections, the contribution of Bourdieu (1990) is helpful at one level, but erroneous at another. Bourdieu suggests that the sociologist should seek to 'objectify' the intellectual and practical means whereby social phenomena are themselves objectified. In principle, that would mean treating seriously all of the aspects of epistemic reflexivity we have identified. But he advocates such an approach in order to render the outcomes of research more objectively scientific, implying that the reflexivity effect can be eliminated. We, by contrast argue that sustained attention to reflexivity can enrich the research process and the research culture, while remaining an indelible feature of it. Sustained attention to textual reflexivity, for instance, does not result in a pristine form of representation that transparently reports the world. But it does provide the ethnographer with a renewed sense of authorship, as diverse styles of reportage are consciously available. The potential to shift 'position' or 'standpoint' provides the ethnographer with the opportunity to adopt different perspectives, illuminating phenomena that might otherwise be overlooked. Awareness of disciplinary reflexivity helps the scholar to address the implicit boundaries and paradigms that tacitly inform their own and others' work, opening up new possibilities. Together they constitute a sustained form of phenomenological bracketing, whereby the taken-for-granted features of social life *and* sociological scholarship are made available to consciousness.

The most significant issue here is that reflexivity and reflective practice are not matters of individual action on the part of the practitioner. Here we concur thoroughly with Bourdieu (and others) on the collective, disciplinary nature of the relevant intellectual tasks:

> As epistemic permeability questions the boundaries between and within disciplines, institutions and the social world, the challenge is to design ways for collectively producing knowledge that is a reflexive ethos, without collapsing into individualised therapy, whilst maintaining concern to contribute to the possibilities of transformation of the world in which we work and belong (May & Perry, 2017: 171).

In other words, reflexivity, leading to critical reflection, frames the work of social scientists as intellectuals, in their shared and collective labour of research and analysis.

A recognition of reflexivity in the widest sense implies that the phenomena we describe and write about are constructed through the *work* we do as researchers and authors. A recognition of such constructive decision-making does not, however, mean treating such matters as bogus; the fact that all methods of measurement 'refract' aspects of the social world does not thereby diminish research. The same is as true of research in the natural sciences as it is among the social disciplines. The images that are generated by laboratory or biomedical sciences, for instance, are not direct, transparent representations of phenomena. They are mediated by the available technologies. Dumit's study of Positron Emission Tomography (PET) is a significant case in point (Dumit, 2004). He displays how images of brain activity are generated through complex techniques of enhancement and management. To say that the signals are 'manipulated' in order to render them intelligible is not to cast doubt on them, but to acknowledge that phenomena do not speak for themselves. The same is true of an analysis of the 'aesthetic' work that natural scientists perform in order to render data intelligible—through the use of 'false colour', for instance (Lynch & Edgerton, 1987).

Now it is abundantly clear that at root our treatment of reflexivity means that we employ a constructionist view of knowledge-generation, especially as it relates to work in the social sciences. A perspective grounded in the *social construction of reality*, or some such designation, is often regarded as a device for undermining the validity of such knowledge itself, as it may be regarded as 'mere' construction. That is not our intention, there is nothing 'mere' about it, and it is not a necessary consequence of addressing reflexivity. Rather, we prefer to suggest that such a perspective alerts us to the fact that all knowledge is the outcome of individual and collective work. It is *produced*. We make sense of social and natural phenomena by using the methods, tools and resources to hand. In doing so we exercise professional judgement, guided by the disciplines into which we have been socialised. Understanding those processes is an integral part of the scholarly and scientific enterprise. It in no way diminishes the value of such work. Indeed, wilfully remaining purblind to them betrays the endeavours that go into knowledge production.

Rather than stereotypes of objective 'science' and subjective 'arts', we might more usefully think of social research as a type of *craft* knowledge. The craft worker has a thorough knowledge of the 'apparatus' that she or he uses. The apparatus of craft is the assemblage of knowledge and material resources that are brought to bear. Skilled craft workers have a full understanding of the raw materials (wood, clay, textiles, pigments) and of the tools of the craft. Their apparatus is the entire assemblage of the workshop or studio. They bring to bear knowledge and experience. Their designs are based on precedents and exemplars (their own and others'). Mentors help to enculturate novices into the techniques and expressive possibilities of the craft. Their work is 'objective', in the sense that it engages with the practical possibilities and limitations of their materials and their apparatus. The same is true of practical science. Novice scientists are socialised at the bench, in the laboratory. In the course of practice, they develop tacit skills of laboratory work, and they acquire practical acquaintance with the materials they work with (Delamont et al., 2000). They learn how to 'work' the apparatus and to get their experiments to 'work' in consequence. It is not simply a metaphor to suggest that science is a craft. The same is equally true of social research.

If we allow that sociological or anthropological work is craft, then we must not and do not need to empty it of personal, professional judgement. Judgement and experience are inescapably brought to bear. Like the craft worker's experience, it is grounded in the traditions, genealogies and exemplars of disciplinary fields. The social researcher has to acquire a firm understanding of the apparatus: the relevant methods, techniques and technologies of data production, analysis and representation. Professional judgement is brought to bear, and is developed through practical, concrete engagement with the activities of research itself. Personal, often tacit, knowledge is deployed in bringing together techniques, resources and methods. Craft knowledge informs the evaluation of research outcomes, just as potters judge their pots, just as scientists evaluate the outcomes of experiments or other interventions. Craft in this sense also implies robust and rigorous self-criticism: failed pots are scrapped, unsuccessful techniques are re-evaluated, mistakes are occasions for new learning, and communities of practice provide criteria of success. Discipline and personal experience are thoroughly interlinked. A thorough grasp of reflexivity and its implications inform all decisions of those kinds throughout the conduct and evaluation of social research.

References

Anderson, L. (2006). Analytic autoethnography. *Journal of Contemporary Ethnography, 35*(4), 373–395.

Atkinson, P. (2006). Rescuing autoethnography. *Journal of Contemporary Ethnography, 35*(4), 400–404.

Bourdieu, P. (1990). *The logic of practice*. Polity Press.

Davies, C. A. (2008). *Reflexive ethnography: A guide to researching selves and others* (2nd ed.). Routledge.

Delamont, S. (2009). The only honest thing: Autoethnography, reflexivity and small crises in fieldwork. *Ethnography and Education, 4*(1), 51–63.

Delamont, S., Atkinson, P., & Parry, O. (2000). *The doctoral experience: Success and failure in graduate school*. Falmer.

Dumit, J. (2004). *Picturing personhood: Brainscans and biomedical identity*. Princeton University Press.

Lynch, M., & Edgerton, S. Y., Jr. (1987). Aesthetics and digital image processing: Representational craft in contemporary astronomy. *The Sociological Review, 35*(1 suppl), 184–220.

Madison, D. S. (2006). The dialogic performative in critical ethnography. *Text and Performance Quarterly, 26*(4), 323–324.

May, T., & Perry, B. (2017). *Reflexivity: The essential guide*. Sage.

Myerhoff, B., & Ruby, J. (1982). Introduction. In J. Ruby (Ed.), *A crack in the mirror: Reflexive perspectives in anthropology* (pp. 1–35). University of Pennsylvania Press.

Schön, D. (1983). *The reflective practitioner: How professionals think in action*. Temple Smith.

Stuart, F. (2018). Reflexivity: Introspection, positionality, and the self as research instrument—Toward a model of abductive reflexivity. In C. Jerolmack & S. Khan (Eds.), *Approaches to ethnography: Analysis and representation in participant observation* (pp. 211–237). Oxford University Press.

Index

E. M. Whitaker and P. Atkinson, *Reflexivity in Social Research*,
https://doi.org/10.1007/978-3-030-84095-2

F

G

I

M

N

O

P

Q

R

S

T

The manufacturer's authorised representative in the EU is Springer Nature Customer Service Centre GmbH, Europaplatz 3, 69115 Heidelberg, Germany. If you have any concerns regarding our products, please contact ProductSafety@springernature.com

Printed and bound by CPI Group (UK) Ltd, Croydon, CR0 4YY
30/09/2026
02225475-0001